INTERNATIONAL UNION OF
PURE AND APPLIED CHEMISTRY

COMPENDIUM OF ANALYTICAL NOMENCLATURE

INTERNATIONAL UNION OF
PURE AND APPLIED CHEMISTRY

ANALYTICAL CHEMISTRY DIVISION

COMPENDIUM OF ANALYTICAL NOMENCLATURE

DEFINITIVE RULES 1977

Prepared for publication by

H. M. N. H. IRVING
(Chemistry Dept., University of Leeds, Leeds, UK)

H. FREISER
(Chemistry Dept., University of Arizona, Tucson, USA)

and

T. S. WEST
(Macaulay Institute for Soil Research, Aberdeen, UK)

PERGAMON PRESS

OXFORD · NEW YORK · TORONTO · SYDNEY · PARIS · FRANKFURT

U.K.	Pergamon Press Ltd., Headington Hill Hall, Oxford OX3 0BW, England
U.S.A.	Pergamon Press Inc., Maxwell House, Fairview Park, Elmsford, New York 10523, U.S.A.
CANADA	Pergamon of Canada Ltd., 75 The East Mall, Toronto, Ontario, Canada
AUSTRALIA	Pergamon Press (Aust.) Pty. Ltd., 19a Boundary Street, Rushcutters Bay, N.S.W. 2011, Australia
FRANCE	Pergamon Press SARL, 24 rue des Ecoles, 75240 Paris, Cedex 05, France
FEDERAL REPUBLIC OF GERMANY	Pergamon Press GmbH, 6242 Kronberg–Taunus, Pferdstrasse 1, Federal Republic of Germany

Copyright © 1978 International Union of Pure and Applied Chemistry

All Rights Reserved. No part of this publication may be reproduced, stored in a retrieval system or transmitted in any form or by any means: electronic, electrostatic, magnetic tape, mechanical, photocopying, recording or otherwise, without permission in writing from the copyright holders

First edition 1978

Library of Congress Cataloging in Publication Data

Main entry under title:

Compendium of analytical nomenclature.

(IUPAC publication)
At head of title: International Union of Pure
and Applied Chemistry, Analytical Chemistry Division.
Includes index.
1. Chemistry, Analytic--Nomenclature.
2. Chemistry, Analytic--Terminology, I. Irving,
Harry Munroe Napier Hetherington. II. West,
Thomas Summers. III. Freiser, Henry, 1920-
IV. International Union of Pure and Applied
Chemistry. Analytical Chemistry Division. V. Se-
ries: International Union of Pure and Applied
Chemistry. IUPAC publication.
QD75.3.C65 1977   543'.001'4   77-8949
ISBN 0-08-022008 8 (Hard cover)
      0-08-022347 8 (Flexi cover)

Printed by A. Wheaton & Co. Ltd., Exeter

CONTENTS

0.	PREAMBLE	1
	0.1 Sources of the material incorporated in the compendium	1
	0.2 General principles of nomenclature standardization	2
	0.3 Physical quantities: ways of naming physical quantities	3
	0.4 Scope and treatment of the present compendium	5
1.	RECOMMENDATIONS FOR THE PRESENTATION OF THE RESULTS OF CHEMICAL ANALYSIS	8
	1.1 Introduction	8
	1.2 General terms	8
	1.3 Reliability of results	9
	1.4 Expression of results	9
	1.5 Symbols	10
	1.6 Methods of computation	10
	1.7 Appendix	11
2.	RECOMMENDATIONS FOR TERMINOLOGY TO BE USED WITH PRECISION BALANCES	12
	2.1 Introduction	12
	2.2 Terminology	12
	2.3 Glossary of terms	13
3.	RECOMMENDED NOMENCLATURE FOR SCALES OF WORKING IN ANALYSIS	15
	3.1 Introduction	15
	3.2 Sample weight classification	15
	3.3 Constituent content classification	16
4.	RECOMMENDATIONS ON NOMENCLATURE FOR CONTAMINATION PHENOMENA IN PRECIPITATION FROM AQUEOUS SOLUTION	18
	4.1 Introduction	18
	4.2 Definitions	18
5.	RECOMMENDED NOMENCLATURE FOR AUTOMATIC ANALYSIS	22
	5.1 Introduction	22
	5.2 Recommended terminology	22

CONTENTS

6. RECOMMENDATIONS FOR NOMENCLATURE OF THERMAL ANALYSIS — 24
   - 6.1 General recommendations — 24
   - 6.2 Terminology — 24
   - 6.3 Definitions and conventions — 26

7. RECOMMENDATIONS FOR NOMENCLATURE OF MASS SPECTROMETRY — 28

8. RECOMMENDED NOMENCLATURE FOR TITRIMETRIC ANALYSIS — 35

9. REPORT ON THE STANDARDIZATION OF pH AND RELATED TECHNOLOGY — 44
   - 9.1 Introduction — 44
   - 9.2 Symbols — 44
   - 9.3 Operational definition of pH — 44
   - 9.4 Standards for measurement of pH — 45
   - 9.5 Ionic activity coefficients — 46
   - 9.6 pH of standard solutions — 48
   - 9.7 Approximate interpretation of pH — 48
   - 9.8 The abbreviation pa_H — 48

10. PRACTICAL MEASUREMENT OF pH IN AMPHIPROTIC AND MIXED SOLVENTS — 49
    - 10.1 Introduction — 49
    - 10.2 The operational pH scale — 49
    - 10.3 Interpretation of the measured pH — 49
    - 10.4 Extension to other solvents — 50
    - 10.5 Selection of a pH unit for amphiprotic solvents — 51
    - 10.6 Operational pH scale for amphiprotic solvents — 52
    - 10.7 Reference solutions for values of pH amphiprotic solvents — 52
    - 10.8 Accuracy of practical scales — 53

11. RECOMMENDED SYMBOLS FOR SOLUTION EQUILIBRIA — 54
    - 11.1 Introduction — 54
    - 11.2 General rules — 54
    - 11.3 Complex formation equilibria — 55
    - 11.4 Solubility equilibria — 58
    - 11.5 Liquid-liquid distribution equilibria — 58

12. RECOMMENDED NOMENCLATURE FOR LIQUID-LIQUID DISTRIBUTION — 60

13. RECOMMENDATIONS ON NOMENCLATURE AND PRESENTATION OF DATA IN GAS CHROMATOGRAPHY — 64
    - 13.1 Introduction — 64
    - 13.2 Name of technique — 64
    - 13.3 Apparatus — 64
    - 13.4 Reagents — 65
    - 13.5 Chromatographic records — 66

CONTENTS

13.6	Retention parameters	66
13.7	Recommendations: retention data	69
13.8	Apparatus performance	69
13.9	Discussion	70
13.10	Table of terms	72

14. RECOMMENDATIONS ON NOMENCLATURE FOR CHROMATOGRAPHY 74

14.1	Chromatography	74
14.2	Principal methods	74
14.3	Classification according to phases used	74
14.4	Classification according to mechanisms	75
14.5	Classification according to techniques used	75
14.6	Terms for special techniques	76
14.7	Terms relating to the method in general	76
14.8	Terms relating to the separation process and the apparatus	78
14.9	Terms relating to quantitative evaluation and the theory of chromatography	80
14.10	Appendix and list of symbols	86

15. RECOMMENDATIONS ON ION EXCHANGE NOMENCLATURE 88

| 15.1 | Introduction | 88 |
| 15.2 | Definitions | 88 |

16. NOMENCLATURE, SYMBOLS, UNITS AND THEIR USAGE IN SPECTROCHEMICAL ANALYSIS-I. GENERAL ATOMIC EMISSION SPECTROSCOPY 93

16.1	Foreword	93
16.2	General recommendations and practices	93
16.3	Terms and symbols for physical quantities in general use	96
16.4	Terms, symbols, and units related to radiant energy	97
16.5	Terms and symbols for the description of spectrographic instruments	99
16.6	Terms and symbols related to the analytical procedures	101
16.7	Terms and symbols related to fundamental processes occurring in light excitation sources	103
16.8	Photographic intensity measurements (photographic photometry)	108
	Appendix: Application of the concept of optical conductance	111

17. NOMENCLATURE, SYMBOLS, UNITS AND THEIR USAGE IN SPECTROCHEMICAL ANALYSIS - II. DATA INTERPRETATION 114

| 17.1 | Introduction | 114 |
| 17.2 | General concepts | 114 |

CONTENTS

17.3	Analytical functions and curves	116
17.4	Terms relating to small concentrations	117
17.5	Glossary of terms and symbols used	117

18. **NOMENCLATURE, SYMBOLS, UNITS AND THEIR USAGE IN SPECTROCHEMICAL ANALYSIS — III. ANALYTICAL FLAME SPECTROSCOPY AND ASSOCIATED NON-FLAME PROCEDURES** — 118

18.1	Introduction	118
18.2	Terms and symbols for general quantities and constants	120
18.3	Terms, symbols, and units for the description of the analytical apparatus	120
18.4	Terms and symbols relating to the analytical procedure and the performance of an analysis	130
18.5	Terms, symbols, and units relating to radiant energy, and its interaction with matter	137
18.6	Terms, symbols, and units relating to the gaseous state of matter	143
18.7	Index of terms	145

19. **CLASSIFICATION AND NOMENCLATURE OF ELECTROANALYTICAL TECHNIQUES** — 148

19.1	Introduction	148
19.2	Techniques in which neither the electrical double layer nor any electrode reaction need be considered	152
19.3	Techniques that involve double-layer phenomena but in which any electrode reactions need not be considered	152
19.4	Techniques involving electrode reactions and employing constant excitation signals	153
19.5	Techniques involving electrode reactions and variable excitation signals of large amplitude	156
19.6	Techniques involving electrode reactions and variable excitation signals of small amplitude	160
19.7	Index to Tables 19.2–19.6	163

20. **RECOMMENDATIONS FOR SIGN CONVENTIONS AND PLOTTING OF ELECTROCHEMICAL DATA** — 166

21. **RECOMMENDATIONS FOR NOMENCLATURE OF ION-SELECTIVE ELECTRODES** — 168

21.1	General recommendations	168
21.2	Classification of ion-selective electrodes	171
21.3	Constants and symbols	172

APPENDIX: RECOMMENDATIONS ON THE USAGE OF THE TERMS 'EQUIVALENT' AND 'NORMAL' — 175

INDEX — 187

0. PREAMBLE

0.1 SOURCES OF THE MATERIAL INCORPORATED IN THE COMPENDIUM

During the past twenty years various Commissions within the Analytical Division of the International Union of Pure and Applied Chemistry have been engaged in drawing up reports recommending nomenclature and symbols to be used in many disciplines of analytical chemistry, after a most careful scrutiny of current practice and the fullest possible consultations with acknowledged international experts. Individual reports have been published in *Pure and Applied Chemistry* from time to time and in each case at least eight months have been allowed to elapse after a preliminary and provisional publication which has been given wide circulation in order that informed criticism and other amendments should influence the definitive versions published subsequently. The definitive nomenclature reports thus produced have then been published with the full approval of the Council of the International Union.

The need to consult a substantial number of separate publications has undoubtedly reduced the effectiveness of the Analytical Division's efforts to secure full recognition and widespread adoption of the recommended symbols and nomenclature, and the time has now come to assemble these official IUPAC recommendations in a single volume. This task has presented a number of problems that did not arise in the comparable compendia from other Divisions, e.g., *Nomenclature of Inorganic Chemistry* (the 'Red Book', Second Edition, Butterworths, 1970), *Nomenclature of Organic Chemistry* (the 'Blue Book', Third Edition, Butterworths, 1971), and the *Manual of Symbols and Terminology for Physical Quantities and Units* (the 'Green Book', 1973 Edition, Butterworths, 1975). The problem arises to some extent from the great variety of topics to be included and the difficulties inherent in devising any logical order of presentation. However, a more serious problem arises from the fact that the actual membership of the Commissions and various working parties have changed a great deal over the years and there has been no common policy or general directive regarding the style and treatment of the different projects and it is not surprising that individual publications vary a great deal in the manner in which recommendations have been made and discussed. In some cases precise definitions have been recommended for every word and term commonly in use in a given field, in other instances this has not been attempted at all, and all relevant terms have been incorporated in a continuous and discursive prose treatment from which the meanings of the various terms should become obvious from the context.

Irrespective of the ultimate form taken by individual reports, the same general principles have guided all the Commissions in their deliberations. These are summarised in the following statement, which has been freely adapted from Appendix A to the Report by the

Commission on Spectrochemical and other Optical Procedures recently published in *Pure and Applied Chemistry*, Vol.30, Nos. 3 - 4, pp. 674 - 676 (1972), while incorporating a number of points made in other reports.

0.2 GENERAL PRINCIPLES OF NOMENCLATURE STANDARDIZATION

0.2.1 SELF-EXPLANATORY TERMS

A technical term should, whenever possible, be more or less self-explanatory and thus easy to learn and associate with its concepts. Any precise definition should only employ words in common usage together with technical terms that have previously been defined.

Some terms that appear to satisfy these criteria are buffer, load, peak, or calibration curve. On the other hand electromotive force and electric capacity (now obsolete) are examples of terms that are positively misleading.

0.2.2 FAULTY TERMS

If a term is faulty, but is too firmly established to be discarded, a brief explanatory comment should be included with its definition. If a term is faulty in some important respect an effort should be made to change it, no matter how well established it may be. There is no better example of this than 'electromotive force'.

While 'amperometric titration' is clearly more specific and preferable to 'polarographic titration', for somewhat different reasons the terms 'thermal analysis' and 'thermogravimetric curve' should definitely be used in place of 'thermography' and 'thermogram', terms widely used for special medical techniques of quite a different kind.

The term 'raffinate', formerly used to describe a 'refined product', has become very firmly established in the practice of liquid-liquid extraction (solvent extraction) with an extended and indeed almost opposite meaning. In this and in similar cases it would be merely pedantic to insist on the historical rather than on the contemporary usage.

A special difficulty arises where the recommendations of one Commission conflict seriously with established usage. For example, the *Manual of Symbols and Terminology for Physical Quantities and Units* published by Commission I.1 of the Division of Physical Chemistry recommends the symbol I for current, in agreement with the recommendations of other international organisations. However, in electroanalytical chemistry the symbol i has been used almost exclusively in the literature where, too, the quantity E (potential difference) is almost always called simply 'potential'.

Confusion between 'derivative' and 'differential' methods of analysis have sometimes led to quite wrong designations being applied. In other cases the term 'subtractive' has been introduced in an attempt to avoid any confusion.

0.2.3 NEW TERMS

The question often arises as to the procedure to be adopted in choosing an entirely new term for a new idea or a technique as yet unnamed. There are a number of alternatives.

0.2.3.1 Borrow a word from ordinary language. The resulting term could be ambiguous because of the possible difficulty of determining from the context whether the ordinary meaning of the word or the (new) technical meaning is intended. Moreover, such a term may not be internationally acceptable since it cannot normally be taken over directly into other languages. For example, whereas in the English language there is a definable difference between a programme (e.g., for a concert or public event) and a program (for a computor), this distinction cannot be made in the United States of America where only the spelling 'program' is employed.

While it is true that many existing terms borrowed from ordinary language have proved to be quite satisfactory, it is generally agreed that this way of obtaining a new term should be used with considerable discretion and restraint.

0.2.3.2   Borrow a word from a living foreign language. Willard Gibbs did just this when he took the French word 'ensemble' and gave it a new technical meaning. One serious objection to this, as a general proposition, is that the borrowed term, while good for many people, might be thoroughly unsatisfactory for scientists of the country from which the word was borrowed. Thus the word 'ensemble' in French does not convey Gibbs' idea at all well.

0.2.3.3   Coin a new word by making an arbitrary although euphonious combination of a number of letters of the alphabet. This might serve on occasions, although the resulting new term might be hard to associate with its meaning and hence hard to learn and memorise.

0.2.3.4   Adopt the name of the discoverer of the idea or technique, as in the 'Kjeldahl' or 'Karl Fischer' methods, or in 'Kalousek polarography'. There is, of course, a serious obligation that the name selected should be that of the real discoverer as established by priority of publication.

0.2.3.5   Construct the new technical word from parts taken from a classical language such as Latin or Greek (with Greek words using Latin transliteration to avoid any real ambiguity in spelling). The word, being a new one, is then immediately recognizable as a technical term, and if the component parts have been chosen properly it will also be more or less self-explanatory. Classical Greek is thought by many to be the best language for this purpose. Hybrids of Latin and Greek are rightly frowned upon by the purist, though a number have come into common use.

0.2.3.6   The registered trade mark of a particular company should not be used for a class of instruments.

0.3   PHYSICAL QUANTITIES : WAYS OF NAMING PHYSICAL QUANTITIES

Care must be taken to realise that any physical quantity is the product of a *numerical value* (a pure number) and a *unit*. SI Base Units and the Symbols described and listed in IUPAC's *Manual of Symbols and Terminology for Physical Quantities and Units* should be used consistently. As pointed out above there may be occasions where long established usage in a particular field means that preferred or recommended terms must be accompanied by alternative (less recommended or not recommended) terms to pave the way for complete uniformity at a later stage.

0.3.1   DUPLICATION OF TERMS
There should only be one agreed name for any given analytical concept. For example,

'analytical curve' and 'working curve' are not both needed.

0.3.2 MULTIPLICITY OF MEANINGS

A given term should not have more than one technical meaning in any branch of physical science. For example, if 'density' is accepted as the term for 'mass divided by volume', the same word should not be used as a shortened form for 'optical density' and indeed this term itself should be replaced by the more appropriate term 'absorbance'.

0.3.3 RELATED CONCEPTS

If several concepts are closely related this should be indicated and reflected in the similarity of the names assigned to them. Instruments should be named for the quantities they are to measure.

0.3.3.1 One obvious procedure is to incorporate the name of the most basic of the concepts into the names of those subsidiary to it. Thus when speaking of various standards one should use, for example, 'internal standard', 'secondary standard', etc.

0.3.3.2 Another exceptionally useful way of indicating such relationships is to utilise the conventions regarding suffixes that are now fairly well established. There are at least nine such suffixes:

- *-or*, meaning a device. For example, comparator, reflector, generator, computor, capacitor, titrator. (Other suffixes which also denote devices end in *-er*, as in amplifier, and *-ment*, as in instrument.)
- *-ation*, meaning a process or the result of a process. For example, calibration, ionization, excitation, volatilisation, sublimation, radiation and documentation. (*-ing*, as in precipitating, dissolving, and reducing, also denotes a process).
- *-ance*, meaning a property of a body or device. For example, capacitance, absorbance, and transmittance.
- *-ity*, meaning a property of a substance. For example, radioactivity, conductivity, density, viscosity, resistivity (which is simpler than the older term 'specific resistance', which is in any case faulty since the term 'specific' has now been restricted to the meaning 'divided by mass').
- *-meter*, meaning a measuring device. For example, photometer, ammeter, voltmeter, coulometer, mass spectrometer.
- *-scope*, meaning an optical or viewing device, as in microscope, telescope, and spectroscope.
- *-graph*, meaning a device for producing a record of observations, as in spectrograph and polarograph.
- *-gram*, meaning the record produced by an instrument, as in spectrogram or polarogram. The termination also appears in other words signifying the visual or identifiable record of a process, as in chromatogram.

0.3.4 CONTRASTING CONCEPTS

The names of two diametrically opposite ideas should differ markedly in sound and appearance. An example where this obvious ideal has not been achieved is in the use of the terms 'microscopic' and 'macroscopic' — each consisting of eleven letters with only one of them different: even their pronunciation may lead to confusion. Other outstanding examples are provided by the terms 'radiance' and 'irradiance' in spectrochemical terminology, and 'molarity' and 'molality' in definitions of molecular quantities.

0.3.5 SIMPLICITY

The chosen term should be simple and euphonious and remain unaffected by minor changes in spelling sanctioned by custom in different English-speaking countries. Terms such as 'quantity of heat' and 'quantity of charge' not only lack simplicity but are redundant, for only 'heat' and 'charge' are necessary. 'Sinusoidal function' should give way to 'sinoidal function', and 'coefficient of linear expansion' to 'linear expansivity', which is not only shorter and simpler but utilizes the $-ity$ convention. Problems created by 'program' and 'programme' have been mentioned above.

0.3.6 INTERNATIONALITY

The name for a particular concept should preferably have the same form in a number of the chief languages widely used in scientific communications. This is already true of various technical terms in common use as, for instance, 'energy' and 'entropy' and many of the terms utilizing certain of the prefixes described previously. For the four suffixes, $-or$, $-ation$, $-ance$, and $-ity$, a tabulation of 544 entries from eight different languages showed that one or other of these suffixes was used in 96 per cent of the total. Clearly the proposal to achieve some degree of internationality in terminology is by no means impracticable.

0.4. SCOPE AND TREATMENT OF THE PRESENT COMPENDIUM

The Reports that have been incorporated into the present volume are listed below in the order in which they were published:-

1. Report on the Standardization of pH and Related Terminology. Published in *Pure Appl. Chem.*, Vol. 1, No.1 (1960), pp. 163 - 168.
2. Terminology for Scales of Working in Microchemical Analysis. Published in *Pure Appl. Chem.*, Vol.1, No.1 (1960), pp. 169 - 170.
3. Recommendations for Terminology to be used with Precision Balances. Published in *Pure Appl. Chem.*, Vol. 1, No. 1 (1960), pp. 171 - 175.
4. Preliminary Recommendations on Nomenclature and Presentation of Data in Gas Chromatography. Published in *Pure Appl. Chem.*, Vol. 1, No. 1 (1960), pp. 177 -186.
5. Recommendations on Nomenclature and Presentation of Data in Gas Chromatography. Published in *Pure Appl. Chem.*, Vol. 8 (1964), pp. 553 - 562.
6. Practical Measurements of pH in Amphiprotic and Mixed Solvents. Published in *Pure Appl. Chem.*, Vol. 18, No. 3 (1969), pp. 421 - 425.
7. Recommended Nomenclature for Titrimetric Analysis. Published in *Pure Appl. Chem.*, Vol. 18, No. 3 (1969), pp. 427 - 436.
8. Recommendations for the Presentation of Results of Chemical Analysis. Published in *Pure Appl. Chem.*, Vol. 18, No. 3 (1969), pp. 437 - 442.
9. Recommended Symbols for Solution Equilibria. Published in *Pure Appl. Chem.*, Vol.18,

No. 3 (1969), pp. 457 - 464.

10. Recommended Nomenclature for Liquid-Liquid Distribution. Published in *Pure Appl. Chem.*, Vol. 21, No. 1 (1970), pp. 109 - 114.

11. Recommended Nomenclature for Automatic Analysis. Published in *Pure Appl. Chem.*, Vol. 21, No. 4 (1970), pp. 527 - 531.

12. Recommendations on Ion Exchange Nomenclature. Published in *Pure Appl. Chem.*, Vol. 29, No. 4 (1972), pp. 617 - 624.

13. Nomenclature, Symbols, Units and Their Usage in Spectrochemical Analysis - I. General Atomic Emission Spectroscopy. Published in *Pure Appl. Chem.*, Vol. 30, Nos. 3 - 4 (1972), pp. 651 - 679.

14. Recommendations on Nomenclature for Thermal Analysis (Rules 1972). Published in *Pure Appl. Chem.*, Vol. 37, No. 4 (1974), pp. 439 - 444.

15. Recommendations on Nomenclature for Chromatography. Published in *Pure Appl. Chem.*, Vol. 37, No. 4 (1974), pp. 445 - 462.

16. Recommendations on Nomenclature for Contamination Phenomena in Precipitation from Aqueous Solutions (Rules 1973). Published in *Pure Appl. Chem.*, Vol. 37, No. 4, pp. 463 - 468.

17. Recommendations for Nomenclature of Mass Spectrometry. Published in *Pure Appl. Chem.*, Vol. 37, No. 4 (1974), pp. 469 - 480.

18. Classification and Nomenclature of Electroanalytical Techniques. Published in *Pure Appl. Chem.*, Vol. 45, No. 2 (1976), pp. 81 - 97.

19. Nomenclature, Symbols, Units and Their Usage in Spectrochemical Analysis - II. Data Interpretation. Published in *Pure Appl. Chem.*, Vol. 45, No. 2 (1976), pp. 99 - 103.

20. Nomenclature, Symbols, Units, and Their Usage in Spectrochemical Analysis - III. Analytical Flame Spectroscopy and Associated Non-Flame Procedures. Published in *Pure Appl. Chem.*, Vol. 45, No. 2 (1976), pp. 105 - 123.

21. Recommendations for Sign Conventions and Plotting of Electrochemical Data (Rules 1975). Published in *Pure Appl. Chem.*, Vol. 45, No. 2 (1976), pp. 131 - 134.

22. Recommendations for Nomenclature of Ion-Selective Electrodes (Rules 1975). Published in *Pure Appl. Chem.*, Vol. 48, No. 1 (1976), pp. 127 - 132.

23. Recommendations on Scales of Working in Analysis. *Pure Appl. Chem.*, Vol. 50 (1978).

0.4.1 FORMAT AND PRESENTATION

Several different Commissions were engaged in the preparation of the above Reports and they worked independently. Some of the Reports took much longer than others to produce and finalise. For the convenience of the reader the order of presentation in the present Compendium has been altered considerably from the above chronology by grouping related subjects as far as possible and by taking account of those cases (e.g. Nos. 1 and 23, 4 and 5) where a later publication extends or modifies the earlier Report. References will be made to a few cases where still more recent Reports, as yet not finally approved by IUPAC, will augment what has already been published and which may recommend minor changes in points of detail.

In the complete versions of all these Reports as they appear in *Pure and Applied Chemistry* there are accounts of why the work was undertaken, how it developed, and the names of those involved. Such details have been omitted from the present volume in order to economise in space and to keep the size and cost of the Compendium as small as possible. There have been a few Editorial deletions for the same reason. On the other hand, no attempt has been made to change the style of individual Reports by constraining them within a common format.

The index has been prepared with the maximum amount of cross-referencing so that the reader seeking guidance on any particular point of nomenclature can be directed quickly to the relevant Report or Reports in which it has been discussed.

The appendix on usage of the terms 'Normal' and 'Equivalent', pp. 175 - 186, which was added at the Warsaw General Assembly, is not indexed.

1. RECOMMENDATIONS FOR THE PRESENTATION OF THE RESULTS OF CHEMICAL ANALYSIS

1.1 INTRODUCTION

The value of much published research work on chemical analysis is diminished by the lack of a generally accepted system of reporting numerical results. It is seldom that a research worker is able to plan collaborative work for statistical study or, indeed, to carry out more than a short series of replicate determinations in his own laboratory. In the development of a new analytical method, synthetic samples or standard materials of known (assumed) composition are normally used to test the proposed method and it is further assumed that the materials are homogeneous and that sampling errors are negligible. The results of chemical analysis done under the conditions outlined above are, however, amenable to simple statistical treatment.

The following list is designed to provide the necessary means for reporting results in a standardized form with the intention that, by using these terms and symbols, an author may report his data unambiguously without further explanation of terminology or method of computation. Only if other terms were used would it be necessary for the author to define his meaning.

The proposed list includes both statistical terms, e.g. 'standard deviation', and other non-statistical terms, e.g. 'per cent absolute', which are required by the analytical chemist, but which frequently cause confusion even though they are in common use, because differing meanings are attributed to them. The statistical terms are defined according to the practical requirements of the analytical chemist who is normally only concerned with short series. Only the minimum of requirements has been incorporated in the list, but the same terms may be used in more advanced statistical treatment.

Those who are unfamiliar with the elementary theory of statistics may find it useful to consult an introductory text before using these terms.

1.2 GENERAL TERMS

1.2.1 MEASURED VALUE. The observed value of the weight, volume, meter-reading or other quantity found by the chemist in his analysis of a material.

1.2.2 RESULT. The final value reported for a measured quantity after performing a measuring procedure including all sub-procedures *and* evaluations.

*Based on the approved Recommendations published in *Pure and Applied Chemistry*, Vol. 18, No.3 (1969), pp. 437 - 442.

1.2.3 VARIATE.   The numerical value taken for statistical handlong; it may, for example, be a measured value or result. Symbol: x.

>   *Comment.* In circumstances where 'x' may cause confusion, another symbol, e.g. 'z', may be used.

1.2.4 SERIES.   A number of variates ($x_1, x_2, x_3 \ldots x_n$) equivalent to each other with respect to statistical considerations, e.g. the results of repeated analyses using only one analytical method on a substance which is presumed to be homogeneous.

1.3 RELIABILITY OF RESULTS

1.3.1 ACCURACY.   Accuracy relates to the difference between a result (or mean) and the true value.

1.3.2 PRECISION.   Precision relates to the variations between variates, i.e. the 'scatter' between variates.

1.4 EXPRESSION OF RESULTS

A. *In relation to precision*

The reported values should include items 1.4.01, 1.4.03 or 1.4.04 and 1.4.07.

1.4.01 MEAN (AVERAGE).   The sum of a series of variates divided by the number of variates in the series. Symbol: $\bar{x}$.

1.4.02 DEVIATION.   The difference between a variate and the mean of the series to which it belongs. Symbol: d.

>   *Comment.* The symbol 'e' (French: *écart*) is recommended in place of the commonly used symbols 'd' or 'δ' (Greek: delta) where their use may cause any confusion.

1.4.03 RANGE.   The difference between the highest and the lowest results of measured values in a series.

>   *Comment.* This term should only be used where there are insufficient data available for calculation of standard deviation. The estimation of standard deviation from range-tables under these circumstances is not recommended.

1.4.04 STANDARD DEVIATION.   The square root of the sum of the squares of the deviations between the variates and the mean of the series, divided by one less than the total number in the series. Symbol: s.

>   *Comment.* The term 'Standard error' in place of 'Standard deviation' is not recommended.
>   The number of variates necessary to obtain a meaningful standard deviation may be assessed by reference to Table 1.7 in the Appendix. The symbol 'σ' (Greek: sigma) should not be used in place of 's'.

1.4.05 VARIANCE.   The square of the standard deviation. Symbol: V.

>   *Comment.* This term is not normally reported by the analytical chemist, but is used in the process of statistical calculation.

1.4.06 RELATIVE STANDARD DEVIATION.   The standard deviation divided by the mean of the series. Symbol: s_r.

>   *Comment.* It is recommended that the Relative standard deviation should be reported as a decimal fraction, *not* as a percentage, in order to avoid confusion where results themselves are reported as percentages. The term Coefficient of variation in place of Relative standard deviation is not recommended.

1.4.07 NUMBER OF VARIATES. The total number of variates in the series. Symbol: n.

Comment. This number should *always* be reported.

B. *In relation to accuracy*

1.4.08 ERROR. The value of a result minus the true value.

Comment. Where a result is reported as a percentage this term will, of course, appear as a percentage. In these circumstances, in order to differentiate between this term and 'Percentage error' — see 1.4.09 — it is then permissible to refer to the error as 'Percent absolute'.

1.4.09 PERCENTAGE ERROR. The error expressed as a percentage of the true value.

Comment. The term 'Percentage error' must be quoted in full rather than 'Error' to avoid confusion with the term 'Error' as defined above. The term 'Relative error' should not be used.

1.4.10 BIAS. The mean of the differences, having regard to sign, of the results from the true value. This equals the difference between the mean of a series of results and the true value.

1.4.11 PERCENTAGE RECOVERY. A result expressed as a percentage of the true value.

1.5 SYMBOLS

1.5.01	VARIATE	x
1.5.02	SERIES	$x_1, x_2, x_3 \ldots x_n$
1.5.03	NUMBER IN SERIES	n
1.5.04	SERIES SUM	Σx
1.5.05	MEAN (AVERAGE)	$\bar{x}$
1.5.06	DEVIATION	d
1.5.07	VARIANCE	V
1.5.08	STANDARD DEVIATION	s

1.6 METHODS OF COMPUTATION

A. *From a series of variates at one level of the quantity being determined*

1.6.01 MEAN $\quad \bar{x} = (\Sigma x)/n$

1.6.02 DEVIATION $\quad d = |x - \bar{x}|$

1.6.03 STANDARD DEVIATION
$$s = \{(\Sigma d^2)/(n-1)\}^{1/2}$$
$$= \{[n\Sigma x^2 - (\Sigma x)^2]/n(n-1)\}^{1/2}$$
$$= \{(\Sigma x^2 - [(\Sigma x)^2/n])/(n-1)\}^{1/2}$$

1.6.04 VARIANCE $\quad V = s^2$

1.6.05 RELATIVE STANDARD DEVIATION $\quad s_r = s/\bar{x}$

B. *Computation of pooled standard deviation (symbol: s_g) from more than one series of variates over a range of levels of the quantity being determined*

(i) For each series, having $n_1, n_2, \ldots n_g$ variates respectively, compute the mean and sum the squares of the deviations, $\Sigma d_1^2 + \Sigma d_2^2 + \ldots \Sigma d_g^2$.

(ii) Divide the above sum of squares by the total number of variates $(n_1 + n_2 + \ldots n_g)$ minus the number of series being considered (g).

(iii) Derive the square root of the above quotient.

$$s_g = \left[\frac{\Sigma d_1^2 + \Sigma d_2^2 + \ldots \Sigma d_g^2}{(n_1 + n_2 + \ldots n_g) - g}\right]^{1/2}$$

1.7 APPENDIX

In reporting results in relation to the precision of working, it should be noted that the number of variates 'n' should always be quoted because this allows the reader of a paper to form an opinion of the reported data. This is particularly important where only a standard deviation (s) is recorded. In chemical analysis the distribution of variates about the mean closely follows a normal Gaussian law. The probability of a single value falling within $t.s$ of the mean, $\bar{x}$ of a finite series depends on the number, n, in the series. Table 1.7, based on Gaussian distribution, shows the appropriate value of 't' for different probability levels and values of 'n'.

Table 1.7

Probability level, n	t		
	95%	99%	99.7%
2	12.71	63.66	235
3	4.30	9.92	19.2
4	3.18	5.84	9.22
5	2.78	4.60	6.62
6	2.57	4.03	5.51
7	2.45	3.71	4.90
8	2.37	3.50	4.53
9	2.31	3.36	4.27
10	2.26	3.25	4.09
25	2.06	2.80	3.34
∞	1.96	2.58	3.00

2. RECOMMENDATIONS FOR TERMINOLOGY TO BE USED WITH PRECISION BALANCES*

2.1 INTRODUCTION

When a *load* (within the *capacity* of a *precision balance*) is placed on one of the pans the pointer will be displaced from the *zero point of the scale*. After a number of swings the *deflections* decrease and the pointer settles down at the *rest point*. The *value of a division*, expressed in terms of weight units corresponding to a division of the pointer scale, is the reciprocal of the *sensitivity* and usually varies with the load. The *instrumental indication*, i, is given by multiplying the observed deflection or rest point by the value of a division for the load in question. The *correction of direct weighings* is then achieved by subtracting the *no-load indication*, i_o.

The *precision of a balance* depends upon the *precision of indication* and the *precision of a weighing* depends upon the *method of weighing*, the *readability*, and the *milligram equivalent of readability*, and attention must be paid to the *range of applicability of balances* as a function of capacity and precision.

The terms italicized in the above description are defined below.*

2.2 TERMINOLOGY

2.2.01 LOAD. The total weight acting, after counterbalancing, upon the terminal bearing which carries the object being weighed.

2.2.02 CAPACITY. The maximum safe load claimed by the manufacturer.

2.2.03 PRECISION OF INDICATION. The standard deviation of the instrument indication, i, for a stated load.

2.2.04 VALUE OF A DIVISION (in weight units per division of the pointer scale). This is the reciprocal of the sensitivity and, like the latter, usually varies with the load. It is determined by empirical calibration.

2.2.05 INSTRUMENTAL INDICATION, i. The observed deflection or rest point multiplied by the value of the division for the load in question.

2.2.06 NO-LOAD INDICATION, i_o. The deflection or rest point (no-load reading) multiplied by the value of the division for zero load (rider at zero).

2.2.07 DEFLECTION (in terms of divisions of the pointer scale). The other point of reversal of an ideal (undamped) swing starting at the zero point of the pointer scale. Since the points of reversal of an ideal swing are located symmetrically about the rest

---

* Based on the Report of the Commission on Microchemical Techniques (*Pure Appl. Chem.*, 1, 171 - 175 (1960)) where versions in French and German were included.

point, the deflection is equal to twice the rest point.

2.2.08 REST POINT. The position of the pointer with respect to the pointer scale when the motion of the beam has ceased.

2.2.09 ZERO POINT OF THE SCALE. The rest point of the properly adjusted balance with no load on the pans and the rider (or chain) in the zero position.

2.2.10 SENSITIVITY (for a stated load). The response per unit mass in terms of the pointer scale per unit of mass.

2.2.11 READABILITY. The smallest fraction of a division to which the index scale can be read with ease either by estimation or by the use of a vernier. It should normally be expressed in divisions of the pointer scale.

2.2.12 MILLIGRAM EQUIVALENT OF READABILITY. The product of readability and the value of the scale division (in milligrams per division).

2.2.13 PRECISION OF A BALANCE. The standard deviation of the instrument for a stated load (e.g., l_{20} for a 20 g load). A statement of the procedure, conditions, and experience of the observer should be included.

2.2.14 RANGE OF APPLICABILITY OF A BALANCE. This is a function of its capacity and precision as shown in Table 2.2.14.

TABLE 2.2.14. Range of applicability of a balance as a function of capacity and precision

Type of balance	Capacity/g	Load/mg
Analytical balance	50 - 200	0.01 - 0.05
Microchemical balance	5 - 20	0.001 - 0.005
Assay balance	1 - 5	0.0005 - 0.002

2.2.15 PRECISION OF A WEIGHING. This depends upon the method of weighing and upon the precision of indication, l, for the load in question as shown in Table 2.2.15.

TABLE 2.2.15. Precision of a weighing by various weighing procedures

Procedure	$\pm (l^2 + l_o^2)^{\frac{1}{2}}$
Direct weighing	$1.4 l_o$
For small loads	$1.4 l$
Substitution	$l/1.4$
Transposition	

2.2.16 CORRECTION OF DIRECT WEIGHINGS. Direct weighings are corrected by subtracting the no-load indication. Direct weighings to be used for the determination of the arm ratio must be corrected for the no-load indication.

2.3 GLOSSARY OF TERMS

	English	French	German
2.3.01	Analytical balance	Balance analytique	Analysenwaage
2.3.02	Assay balance	Balance d'essai	Probierwaage
2.3.03	Capacity	Capacité	Tragfähigkeit
2.3.04	Correction of direct weighing	Correction de la pesée directe	Berichtigung von Proportionalwägungen

TERMINOLOGY TO BE USED WITH PRECISION BALANCES

	English	French	German
2.3.05	Deflection	Déflexion	Ausschlag
2.3.06	Instrumental indication	Indication de l'instrument	Instrumentanzeige
2.3.07	Load	Charge	Belastung
2.3.08	Microchemical balance	Balance microchemique	Mikrochemische Waage
2.3.09	Milligram equivalent of readability	Sensibilité de la lecture en milligrammes equivalent	Milligrammäquivalent der Ablesbarkeit
2.3.10	No-load indication	Indication (valeur) à vide	Leeranzeige
2.3.11	Precision	Précision	Präzision
2.3.12	Precision of a balance	Précision de la pesée	Präzision der Wägung
2.3.13	Range of applicability of a balance	Limite d'application de la balance	Anwendungsbereich der Waage
2.3.14	Range of applicability of a balance	Limite d'application de la balance	Anwendungsbereich der Waage
2.3.15	Readability	Appréciation de la lecture	Ablesbarkeit
2.3.16	Rest point	Position de repos	Ruhepunkt
2.3.17	Sensitivity	Sensibilité	Empfindlichkeit
2.3.18	Value of a division	Valeur d'une division	Masse pro Skalenteil
2.3.19	Weighing procedures	Méthodes	Wägeverfahren
2.3.20	Direct weighing	Pesée directe	Proportionalwägung
2.3.21	For small loads	Pour charge faible	Bei kleinen Belastungen
2.3.22	Substitution	Substitution	Substitutionswägung
2.3.23	Transposition	Transposition	Gauss'sche Doppelwägung
2.3.24	Zero point of the scale	Zéro de l'échelle	Nullpunkt der Zeigerskala

3. RECOMMENDED NOMENCLATURE FOR SCALES OF WORKING IN ANALYSIS

3.1 INTRODUCTION

"Scales of working" in analysis primarily implies the size of the sample (test portion) taken. To the extent that there is a choice, sample size,'S', is determined by the method (specifically *procedure*) applied, the relative content of constituent, 'C', and other factors such as the precision required. It is desirable to have a scheme for the classification of analytical methods based on the magnitudes of S and C. It is proposed that methods (procedures) be described and classified, from the standpoint of the scale of working, with the aid of a bipartite designation:

Sample size (weight) - Constituent content (e.g. in percent or ppm)

which can be extended to liquid and gaseous samples. When the ranges of these two variables are given, the range of the absolute quantity, 'Q' of the constituent is of course fixed.

Methods can be classified with any desired degree of fineness on this basis and their fields represented in a Cartesian plot, in which sample weights are plotted as abscissae and relative contents as ordinates. Convenient units are g for S and % or ppm for C. Because of the wide ranges that need to be covered, a double logarithmic plot of S and C is required (Fig.3.1). Diagonal lines in the Figure represent the absolute amounts, Q, of a particular constituent.

Although numbers alone are sufficient, and indeed necessary, for the precise designation and classification of methods, it is convenient, both in written and oral communication, to designate the size ranges of S and C by suitable terms. The use of words is especially convenient when approximate ranges are to be indicated.

3.2 SAMPLE WEIGHT CLASSIFICATION (S)

Sample sizes can be classified as gram (1-10 g), decigram (0.1-1 g), centigram (0.01-0.1 g), milligram (0.001-0.01 g), microgram (10^{-6}-10^{-3} g), nanogram (10^{-9}-10^{-6} g), picogram (10^{-12}-10^{-9} g), femtogram (10^{-15}-10^{-12} g), etc. See A, Fig.3.1.

Macro, semimicro, and *micro* have been used for many years to indicate sample sizes and therewith the scale of analytical operations. Such terms serve a useful purpose and are worth retaining if some measure of agreement on their meaning can be obtained.

---

* These approved Recommendations will be published in *Pure Appl. Chem.*, Vol. 50 (1978). They are an extension of the approved nomenclature published in *Pure Appl. Chem.*, Vol.1 (1960), p.143, and incorporate the provisional recommendations published in February in the IUPAC *Information Bulletin,* No.18.

Quite generally, a macro sample is considered to be one weighing more than 0.1 g. An upper limit is not specified, but most methods, considered macro, call for samples in the range 0.1-1 g. The term semimicro is an unfortunate one in that it does not mean half micro, but *larger* than micro. For this reason the term *meso* is preferred to semimicro. A meso sample (semimicro) may then logically be taken as falling in the range 0.1-0.01 g. Samples in the range $10^{-3}-10^{-4}$ may be called submicro, and those below 10^{-4} g, ultramicro, with no lower limit specified for the latter class; see B, in Fig.3.1.

3.3 CONSTITUENT CONTENT CLASSIFICATION (C)

The terms *major*, *minor*, and *trace* may be used to indicate a broad classification of constituents on the basis of their relative contents, as follows:

Major constituent	100 - 1%
Minor constituent	1 - 0.1%
Trace constituent	<0.01% (<100 ppm)

There are good reasons (historical and practical) for setting the upper limit of *trace* at 100 ppm and it has been, until now, advantageous to set no lower limit, so that anything below 100 ppm has been considered as a trace. However, advances in analytical technology now suggest that the trace range should be further sub-divided, taking 100 ppm as the upper limit as follows:

Trace	$10^2 - 10^{-4}$
Microtrace	$10^{-4} - 10^{-7}$
Nanotrace	$10^{-7} - 10^{-10}$
Picotrace	$10^{-10} - 10^{-13}$

In Microanalysis an 'S' classification is normally used, i.e. the analyst is concerned with the smallness or size of his sample and not so much with the relative concentration of the constituent to be determined. Frequently indeed in microanalysis and in submicro or ultramicro analysis the sought constituent is a major one, i.e. C >1%.

In Trace analysis, on the other hand, the value of 'C' is of paramount importance and usually the value of 'S', the sample size, is a minor consideration. Consequently 'C' may be $10^{-2}-10^{-5}$ ppm and 'S' may be 1 - 100 g.

There are, however, occasions when constituent 'C' may lie in the ppm or sub-ppm level, a typical trace problem, but where the sample size 'S' may only be 100 μg, i.e. a true microanalysis type of problem. In such instances it is felt that an S or C classification is not sufficient and that an S/C one may be necessary.

It is, therefore, proposed that the term *Ultra-trace* (i.e. Ultramicro-trace) be reserved for such analyses. This term could be used generally to describe the whole area of trace analysis using micro sized samples, but could where desired be more precisely specified as follows:

Ultra-trace Analysis, i.e. $S \leq 10^{-4}$ g; $C \leq 100$ ppm (0.01%)

For larger sized samples, similarly one would have the general terms:

Sub-trace Analysis   i.e.   S $10^{-3} - 10^{-4}$ g; C $\leq$ 100 ppm (0.01%)

Micro-trace Analysis i.e.   S $10^{-2} - 10^{-3}$ g; C $\leq$ 100 ppm (0.01%)

Meso-trace Analysis   i.e.   S $10^{-1} - 10^{-2}$ g; C $\leq$ 100 ppm (0.01%)

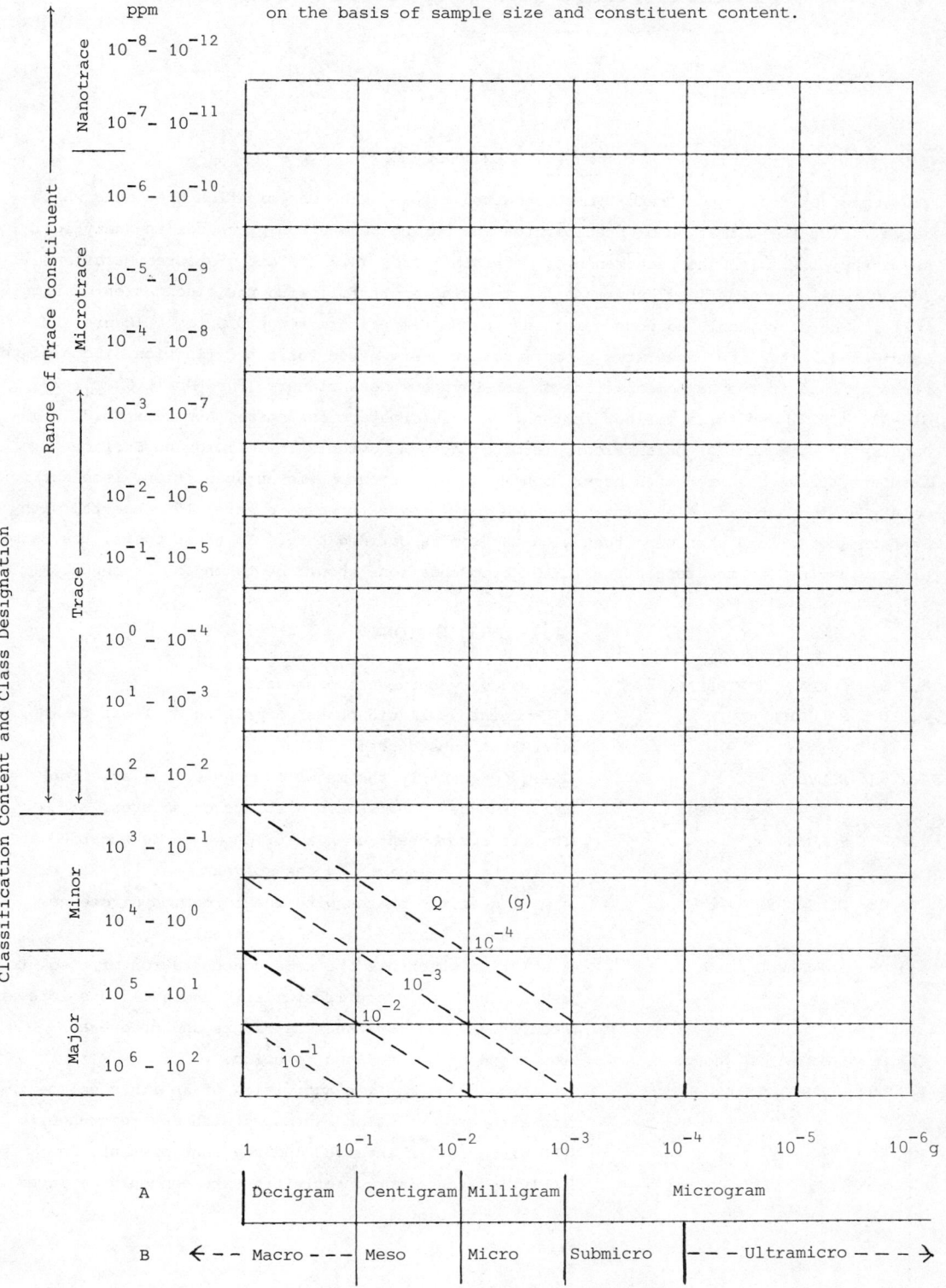

Fig. 3.1 Classification of analytical methods (and procedures) on the basis of sample size and constituent content.

4. RECOMMENDATIONS ON NOMENCLATURE FOR CONTAMINATION PHENOMENA IN PRECIPITATION FROM AQUEOUS SOLUTION*

4.1 INTRODUCTION

The terms defined in this report are connected mainly with contamination phenomena which occur during the formation of precipitates and during separation processes in analytical chemistry, i.e. with the phenomenon of *Coprecipitation* (4.2.34) and *Postprecipitation* (4.2.35) which are defined subsequently in this report following the elucidation of simpler terms. Therefore, only aqueous solutions in the temperature range $0°C$ to $100°C$ are considered. In a few cases, two synonymous terms are given for one definition. Some of the terms defined in the Recommendations on Ion-Exchange Nomenclature (*Pure Appl. Chem.*, 29, 619-624 (1972)) and terms defined in the IUPAC *Information Bulletin*, Appendix No.3, January 1970, Manual of Definitions, Terminology, and Symbols in Colloid and Surface Chemistry, have been repeated here; others (e.g. various terms on absorption, iscelectric point, co-ion, etc.) were omitted because they seemed to be more strongly connected with Surface and Colloid phenomena than with the topics defined here. In these cases, the terms already defined in the respective IUPAC Recommendations should be taken into account.

4.2 DEFINITIONS

4.2.01	SURFACE (INTERFACE)	The boundary between two phases.
4.2.02	SOLUTION	A homogeneous liquid phase comprising at least two different substances.
4.2.03	SOLVENT	A liquid (usually the major component of a solution) which is used to dissolve a solute or solutes.
4.2.04	SOLUTE	The minor component of a solution which is regarded as having been dissolved by the solvent.
4.2.05	DISSOLUTION	A mixing of two phases with the formation of one new homogeneous phase (i.e. the solution).
4.2.06	SATURATED SOLUTION	A solution which has the same concentration of a solute as one that is in equilibrium with undissolved solute at specified values of the temperature and pressure.
4.2.07	SATURATION (noun)	The state of a saturated solution.
4.2.08	SOLUBILITY PRODUCT	The product of the ion activities of an ionic solute in its saturated solution expressed with due reference to the dissociation involved and the ions present. In analytical chemistry, activities are commonly replaced by concentrations.

---

* Based on the approved Recommendations published in *Pure and Applied Chemistry*, Vol. 37, No. 4 (1974), pp. 463 - 468.

4.2.09	SOLUBILITY	The analytical concentration of a solute in a saturated solution. The analytical concentration includes those of all the species formed by the dissolved substance in the solution. Numerical data for solubility always have to be defined in relation to the values of the temperature, pressure, and concentrations of other dissolved substances.
4.2.10	SUPERSATURATED SOLUTION	A labile or metastable solution which contains a higher concentration of solute than a saturated solution at the same temperature and pressure.
4.2.11	SUPERSATURATION (noun)	The state of a supersaturated solution.
4.2.12	MACRO-COMPONENT	A substance dissolved in a solution at such concentration that it may be precipitated by the addition of suitable reagents.
4.2.13	MICRO-COMPONENT	A substance present in solution which is not normally precipitated because of its low concentration or high solubility.
4.2.14	NUCLEUS	The smallest solid phase aggregate of atoms, molecules, or ions which is formed during a precipitation and which is capable of spontaneous growth.
4.2.15	NUCLEATION	The process by which nuclei are formed in a solution. The condensation of a single chemical compound is called homogeneous nucleation. The simultaneous condensation of more than one compound is called simultaneous nucleation. The condensation of a compound on a foreign substance is called heterogeneous nucleation.
4.2.16	RATE OF NUCLEATION	The number of nuclei formed in unit time per unit volume of the liquid phase.
4.2.17	PRECIPITATE (noun)	A solid phase formed within a liquid phase.
	(verb)	To produce a precipitate.
4.2.18	PRECIPITATION	The formation of a precipitate.
4.2.19	PRECIPITATION FROM HOMOGENEOUS SOLUTION (PFHS)	The formation of a precipitate which is generated homogeneously and, generally, slowly by a precipitating agent within a solution.
4.2.20	CONTAMINATION (OF A PRECIPITATE)	The presence of minor amounts of at least one chemically different species in a precipitate.
4.2.21	COLLECTION	The planned removal from solution of a micro- or macro-component by the intentional formation of a contaminated host precipitate, or by the adsorption or entrapment of the component by an added solid.
4.2.22	COLLECTOR (SCAVENGER)	A solid substance added to or formed within a solution to collect a micro- or macro- component.
4.2.23	AGGREGATE	A group of particles held randomly together.
4.2.24	AGGREGATION	The formation of an aggregate.
4.2.25	COAGULATION (FLOCCULATION)	The formation and growth of aggregates ultimately leading to phase separation on a macroscopic scale.
4.2.26	AGGLOMERATION	The formation and growth of aggregates ultimately leading to phase separation by the formation of precipitates

		of larger than colloidal size.
4.2.27	ADSORPTION	An increase in the concentration of a dissolved substance at the interface of a condensed and a liquid phase due to the operation of surface forces.
4.2.28	ABSORBENT	A condensed phase at the surface of which adsorption may occur.
4.2.29	ADSORBATE	The material accumulated on the surface of an adsorbent by the process of adsorption.
4.2.30	OCCLUSION (MOLECULAR)	The process of incorporation of foreign substances as molecular species within the precipitates as they are formed.
4.2.31	MECHANICAL ENTRAPMENT	(a) The process of random incorporation of comparatively small quantities of other phases (e.g. water, dust particles, etc.) in the bulk of a precipitate during its formation.
		(b) The deliberate capture of small quantities of such phases by the deliberate addition of solids to a liquid phase.
	Comment.	The term 'Inclusion' is not recommended.
4.2.32	MIXED CRYSTAL OR SOLID SOLUTION	A crystal which contains a second constituent which fits into and is distributed in the lattice of the host crystal. (The use of 'solid solution' for amorphous materials is not recommended.)
4.2.33	LAWS OF DISTRIBUTION	During the formation of a mixed crystal from a solution containing two components 'A' and 'B', the latter may be distributed according to the equation

$$K_{A,B} = b(a_o - a)/a(b_o - b)$$

In this homogeneous distribution, a_o and b_o are the respective concentrations in the solution before crystallization and a and b are the respective concentrations in the solution after crystallization. $K_{A,B}$ is usually called the separation factor. The term *homogeneous distribution coefficient* is not recommended. Alternatively the distribution of the micro-component may follow the equation of Doerner and Hoskins

$$\ln(a_o/a) = \lambda \ln(b_o/b)$$

(logarithmic distribution) where λ is usually called the *Logarithmic distribution coefficient*, the meaning of the other symbols remaining the same as above.

Exactly homogeneous or logarithmic distributions are extreme cases and are very seldom encountered.

4.2.34	COPRECIPITATION	The simultaneous precipitation of a normally soluble component with a macro-component from the same solution by the formation of mixed crystals, by adsorption, occlusion or mechanical entrapment.

4.2.35	POSTPRECIPITATION	The subsequent precipitation of a chemically different species upon the surface of an initial precipitate usually, but not necessarily, including a common ion.
4.2.36	REPRECIPITATION	The planned repetition of a precipitation in order to remove chemically different species from a previous precipitate or to improve its stoicheiometry.
4.2.37	AGEING	The time-dependent change of the properties of a precipitate, e.g. loss of water, growth of crystals, recrystallization, decrease of the specific surface, loss of coprecipitated substances or improvement of the filtering properties. The process of ageing is very often promoted by maintaining the precipitate and precipitation medium together at elevated temperatures for a period of time. The terms 'chemical', 'physical' and 'thermal' ageing may be used in cases in which some of the (usually combined) effects named above are to be emphasized specifically.
4.2.38	OSTWALD RIPENING	The growth of larger crystals from those of smaller size which have a higher solubility than the larger ones.

5. RECOMMENDED NOMENCLATURE FOR AUTOMATIC ANALYSIS*

5.1 INTRODUCTION

The term *automation* is very commonly used at the present time and is often understood to mean the replacement, partial or complete, of a manual operation, or sequence of operations during the course of an analysis. This wide use of the term robs it of any precise meaning and, in many cases, makes it merely a synonym for mechanization or instrumentation.

The list of terms which follows is proposed so that the three concepts may be distinguished. All are concerned with the diminution of human intervention in analytical procedures; mechanization is concerned with the production of *movement*, instrumentation with the production and relaying of *information*, and automation with the use of systems in which an element of non-human *decision* is incorporated.

A distinction is proposed between the words automatic, as used in normal English, and automated, which implies more than does automatic: viz., that feed-back is applied in an automated system as a result of which the performance of the system is modified. If this distinction between the two ideas is accepted, two sets of words are required, for which the following are suggested; automatize (already in the dictionary though little used), automatic, automatization and automate, automated, automation.

Automation is a term coined for use in fields other than chemistry. However, the authors have not found a satisfactory definition of the word, and have therefore endeavoured to fill this gap as far as chemistry is concerned; they hope that in doing so they have not run counter to usage accepted in other disciplines.

5.2 RECOMMENDED TERMINOLOGY

5.2.01	MECHANISM	A combination of parts, of which one at least is movable, capable of producing an effect.
5.2.02	MACHINE	A device, including one or several mechanisms, that can be made to produce useful work.
5.2.03	MECHANICAL	Relating to or concerned with machines or mechanisms.
5.2.04	MECHANIZATION	The use of mechanical devices to replace, refine, extend or supplement human effort.
5.2.05	MECHANIZE	To equip with or to use mechanical devices.
5.2.06	INSTRUMENT (noun)	A device, used for observing, measuring or communicating

* Based on the approved Recommendations published in *Pure and Applied Chemistry*, Vol. 21 (1970), pp. 527 - 531.

the state of a quality, which replaces, refines, extends, or supplements human faculties.

Note. An instrument may include one or more mechanisms and / or machines.

5.2.07 INSTRUMENTATION The use of instruments.

Note. This term may be used to describe an assembly of instruments used for a particular purpose, but should not be used as a collective noun for instruments in general.

5.2.08 INSTRUMENTAL Relating to or concerned with instruments.

5.2.09 INSTRUMENT (verb) To equip with, or to use, instruments.

Note. This term is preferred, in English, to the synonymous terms *instrumentate* and *instrumentalize*.

5.2.10 PROGRAMME (verb) To equip apparatus with commanding devices that require the performance of given operations (discrete, sequential, or continuous) in single or repeated cycles.

Note. A programmed apparatus contains no self-adjusting devices and thus cannot vary its performance without human intervention.

Note. The spelling '*program*' is becoming widely used.

5.2.11 FEED-BACK SYSTEM A combination of a sensing and a commanding device which can modify the performance of a given act.

Notes. (1). Feed-back control may be continuous or discontinuous.
(2). A feed-back system is an instrumental device.

5.2.12 AUTOMATIC Having self-acting or regulating devices that cause certain required acts to be performed at given points in an operation, without human intervention.

Note. This term may be used to describe mechanical or instrumental devices which perform in accordance with a *manually* preset set of conditions.

5.2.13 AUTOMATIZE (verb) To make automatic. The corresponding noun is *automatization*.

5.2.14 AUTOMATE (verb) To replace human manipulative effort and faculties in the performance of a given process by mechanical and instrumental devices which are regulated by feed-back of information, so that the apparatus is self-monitoring or self-adjusting. The corresponding adjective is *automated*.

5.2.15 AUTOMATION The use of combinations of mechanical and instrumental devices to replace, refine, extend, or supplement human effort and faculties in the performance of a given process, in which at least one major operation is controlled, without human intervention, by a feed-back system.

6. RECOMMENDATIONS FOR NOMENCLATURE OF THERMAL ANALYSIS

6.1 GENERAL RECOMMENDATIONS

6.1.01. Thermal analysis and not 'thermography' should be the accepted name in English, since the latter has at least two other meanings in this language, the major one being medical (*Sci. Progr. London*, Vol. 55 (1967), p.167). The adjective should then be thermoanalytical (cf. physical chemistry and physicochemical): the term 'thermoanalysis' is not supported (on the same logical basis).

6.1.02. Differential should be the adjectival form of difference; derivative should be used for the first derivative (mathematical) of any curve.

6.1.03. The term 'analysis' should be avoided as far as possible since the methods considered do not comprise analysis as generally understood chemically: terms such as differential thermal analysis are too widely accepted, however, to be changed.

6.1.04. The term curve is preferred to 'thermogram' for the following reasons:
   (i) 'Thermogram' is used for the results obtained by the medical technique of thermography.
   (ii) If applied to certain curves (e.g. thermogravimetric curves), 'thermogram' would not be consistent with the dictionary definition.
   (iii) For clarity there would have to be frequent use of terms such as differential thermogram, thermogravimetric thermogram, etc., which are not only cumbersome but also confusing.

6.1.05. In multiple techniques, *simultaneous* should be used for the application of two or more techniques to the same sample at the same time: *combined* would then indicate the use of separate samples for each technique.

6.1.06. *Thermal decomposition* and similar terms are under consideration.

6.2 TERMINOLOGY

Acceptable names and abbreviations are listed in Table 6.2 together with the names which were for various reasons rejected. Consideration has been deferred of techniques not yet

---

* Based upon the approved Recommendations for Nomenclature published in *Pure and Applied Chemistry*, Vol. 37, No.4 (1974), pp. 439 - 444, which fully endorses the recommendations made by the Nomenclature Committee of the International Confederation for Thermal Analysis and published in *Talanta*, Vol. 16 (1969), p. 1227 , et seq.

extensively employed and on matters of nomenclature in borderline techniques such as thermometric titrimetry or calorimetry.

TABLE 6.2 Recommended terminology

	Acceptable name	Acceptable abbreviation *	Rejected name(s)
	General		
6.2.01	Thermal analysis		Thermography, Thermoanalysis
	Methods associated with weight change		
	(i) Static		
6.2.02	Isobaric weight-change determination		
6.2.03	Isothermal weight-change determination		Isothermal thermogravimetric analysis
	(ii) Dynamic		
6.2.04	Thermogravimetry	TG	Thermogravimetric analysis Dynamic thermogravimetric analysis
6.2.05	Derivative thermogravimetry	DTG	Differential thermogravimetry Differential thermogravimetric analysis Derivative thermogravimetric analysis
	Methods associated with energy change		
6.2.06	Heating curves †		Thermal analysis
6.2.07	Heating-rate curves †		Derivative thermal analysis
6.2.08	Inverse heating-rate curves †		
6.2.09	Differential thermal analysis	DTA	Dynamic differential calorimetry
6.2.10	Derivative differential thermal analysis		
6.2.11	Differential scanning calorimetry	DSC	
	Methods associated with evolved volatiles		
6.2.12	Evolved gas detection	EGD	Effluent gas detection
6.2.13	Evolved gas analysis ‡	EGA	Effluent gas analysis Thermovaporimetric analysis
	Methods associated with dimensional change		
6.2.14	Dilatometry		
6.2.15	Derivative dilatometry		
6.2.16	Differential dilatometry		
	Multiple techniques		
6.2.17	Simultaneous TG and DTA, etc.		DATA (Differential and thermogravimetric analysis) Derivatography Derivatographic analysis

* Acceptable abbreviations should be in capital letters without full stops, and should be kept to the minimum to avoid confusion. They should be adopted internationally, irrespective of language.

† When determinations are performed during the cooling cycle these become *Cooling curves*, *Cooling-rate curves*, and *Inverse cooling-rate curves* respectively.

‡ The method of analysis should be clearly stated and abbreviations such as MTA (for mass-spectrometric thermal analysis) and MDTA (mass spectrometry and differential thermal analysis) avoided.

6.3 DEFINITIONS AND CONVENTIONS

General

6.3.01 THERMAL ANALYSIS

A general term covering a group of related techniques whereby the dependence of the parameters of any physical property of a substance on temperature is measured.

6.3.02 METHODS ASSOCIATED WITH WEIGHT CHANGE.

Static

6.3.02.1 ISOBARIC WEIGHT-CHANGE DETERMINATION. A technique for obtaining a record of the equilibrium weight of a substance as a function of temperature (T) at a constant partial pressure of the volatile product or products.

The record is the isobaric weight-change curve: it is normal to plot weight on the ordinate with weight decreasing downwards and T on the abscissa increasing from left to right.

6.3.02.2 ISOTHERMAL WEIGHT-CHANGE DETERMINATION. A technique for obtaining a record of the dependence of the weight of a substance on time (t) at a constant temperature.

The record is the isothermal weight-change curve: it is normal to plot weight on the ordinate with weight decreasing downwards and t on the abscissa increasing from left to right.

Dynamic

6.3.02.3 THERMOGRAVIMETRY (TG). A technique whereby the weight of a substance, in an environment heated or cooled at a controlled rate, is recorded as a function of time or temperature.

The record is the thermogravimetric or TG curve: the weight should be plotted on the ordinate with weight decreasing downwards and t or T on the abscissa increasing from left to right.

6.3.02.4 DERIVATIVE THERMOGRAVIMETRY (DTG). A technique yielding the first derivative of the thermogravimetric curve with respect to either time or temperature.

The curve is the derivative thermogravimetric or DTG curve: the derivative should be plotted on the ordinate and weight losses downwards and t or T on the abscissa increasing from left to right.

6.3.03 METHODS ASSOCIATED WITH ENERGY CHANGE

6.3.03.1 HEATING CURVES. These are records of the temperature of a substance plotted against time, in an environment heated at a controlled rate.

T should be plotted on the ordinate increasing upwards and t on the abscissa increasing from left to right.

6.3.03.2 HEATING-RATE CURVES. These are records of the first derivative of the heating curve with respect to time (i.e. dT/dt) plotted against time or temperature.

The function dT/dt should be plotted on the ordinate and t or T on the abscissa increasing from left to right.

6.3.03.3 INVERSE HEATING-RATE CURVES. These are records of the first derivative of the heating curve with respect to temperature (i.e. dt/dT) plotted against either temperature or time.

The function dt/dT should be plotted on the ordinate and t or T on the abscissa increasing from left to right.

6.3.03.4 DIFFERENTIAL THERMAL ANALYSIS (DTA). A technique of recording the difference in temperature between a substance and a reference material against either time or temperature as the two specimens are subjected to identical temperature regimes in an environment heated or cooled at a controlled rate.

The record is the differential thermal or DTA curve: the temperature difference (T) should be plotted on the ordinate with endothermic reactions downwards and t or T on the abscissa increasing from left to right.

6.3.03.5 DIFFERENTIAL SCANNING CALORIMETRY (DSC). A technique of recording the energy necessary to establish zero temperature difference between a substance and a reference material as the two specimens are subjected to identical temperature regimes in an environment heated or cooled at a controlled rate.

The record is the DSC curve: it represents the amount of heat applied per unit time as ordinate against either t or T as abscissa.

6.3.04. METHODS ASSOCIATED WITH EVOLVED VOLATILES.

6.3.04.1 EVOLVED GAS DETECTION (EGD). This term covers any technique of detecting whether or not a volatile product is formed during thermal analysis.

6.3.04.2 EVOLVED GAS ANALYSIS (EGA). A technique of determining the nature and/or amount of volatile product or products formed during a thermal analysis.

6.3.05 METHODS ASSOCIATED WITH DIMENSIONAL CHANGE

6.3.05.1 DILATOMETRY. A technique whereby changes in dimension(s) of a substance are measured as a function of temperature. The record is the dilatometric curve.

6.3.05.2 DERIVATIVE DILATOMETRY: DIFFERENTIAL DILATOMETRY. These terms carry the connotations given in 6.1.02 above.

6.3.06 MULTIPLE TECHNIQUES

This term covers simultaneous DTA and TG, etc., and definitions follow from the above.

7. RECOMMENDATIONS FOR NOMENCLATURE OF MASS SPECTROMETRY*

7.01 MASS SPECTROMETER

An instrument in which ions are separated according to the quotient mass/charge, and in which the ions are measured electrically.

Note: This term should also be used when a scintillation detector is employed.

7.02 MASS SPECTROGRAPH

An instrument in which beams of ions are separated according to the quotient mass/charge, and in which the deflection and intensity of the beams are recorded directly on photographic plate or film.

7.03 MASS SPECTROSCOPE

A term which may refer to either a mass spectrometer or a mass spectrograph.

7.04 MASS SPECTROMETRY

The branch of science dealing with all aspects of mass spectroscopes and the results obtained from these instruments.

Note: The term '*Mass Spectroscopy*' seems preferable here, but 'mass spectrometry' has become widely used.

7.05 SINGLE-FOCUSING MASS SPECTROMETER

An instrument in which an ion beam with a given value of mass/charge is brought to a focus although the initial directions of the ions diverge.

7.06 DOUBLE-FOCUSING MASS SPECTROMETER

An instrument which uses both direction and velocity focusing, and therefore an ion beam of a given mass/charge is brought to a focus when the ion beam is initially diverging and contains ions of the same mass and charge with different kinetic energies. The ion beam is measured electrically.

7.07 DOUBLE-FOCUSING MASS SPECTROGRAPH

An instrument which uses both direction and velocity focusing, and therefore an ion beam initially diverging in direction and containing ions of different kinetic energies is separated into beams according to the quotient mass/charge, these beams being focused onto a photographic plate or film.

---

* Based on the approved Recommendations published in *Pure and Applied Chemistry*, Vol. 37, No. 4 (1974) pp. 469 -480, in which particular attention was paid to definitions already proposed by the Fachnormen-ausschuss Vacuumtechnik in Deutschen Normenausschuss and in the Editorial Review on Nomenclature in *Organic Mass Spectrometry*, $\underline{2}$, 249 (1969).

7.08 MAGNETIC DEFLECTION

The deflection of an ion beam as a result of the motion of the ions in a magnetic field. Generally the direction of motion of the ions is at right angles to the direction of the magnetic field, and the motion is uniform.

7.09 RADIAL ELECTROSTATIC FIELD ANALYSER

An arrangement of two conducting sheets forming a capacitor and giving a radial electrostatic field which is used to deflect and focus ion beams of different energies. The capacitor may be cylindrical, spherical, or toroidal.

7.10 NIER-JOHNSON GEOMETRY

An arrangement for a double-focusing mass spectrometer in which a deflection of $\pi/2$ radians in a radial electrostatic field analyser is followed by a magnetic deflection of $\pi/3$ radians. The electrostatic analyser uses a symmetrical object-image arrangement and the magnetic analyser is used asymmetrically.

7.11 MATTAUCH-HERZOG GEOMETRY

An arrangement for a double-focusing mass spectrograph in which a deflection of $\pi/4\sqrt{2}$ radians in a radial electrostatic field is followed by a magnetic deflection of $\pi/2$ radians.

7.12 π RADIAN MAGNETIC FIELD ANALYSER

An arrangement in which an ion beam is deflected magnetically through π radians.

7.13 $\pi/2$ RADIAN MAGNETIC FIELD ANALYSER

An arrangement in which an ion beam is deflected magnetically through $\pi/2$ radians.

7.14 $\pi/3$ RADIAN MAGNETIC FIELD ANALYSER

An arrangement by which an ion beam is deflected magnetically through $\pi/3$ radians.

7.15 QUADRUPOLE MASS ANALYSER

An arrangement in which ions with a desired quotient of mass/charge are made to describe a stable path under the effect of a static and a high-frequency electric quadrupole field, and are then measured. Ions with a different mass/charge are separated from the measured ions because of their unstable paths.

7.16 TIME-OF-FLIGHT MASS SPECTROMETER

An arrangement using the fact that ions of different mass/charge need different times to travel through a certain distance ina field-free region after they have all been initially given the same kinetic energy, or the same impulse.

7.17 CYCLOTRON RESONANCE MASS SPECTROMETER

A high-frequency mass spectrometer in which the ions to be detected, with a selected value of the quotient mass/charge, absorb maximum energy through the effect of a high-frequency field and a constant magnetic field perpendicular to the electric field. Maximum energy is gained by the ions which satisfy the cyclotron resonance condition and as a result they are separated from ions of different mass/charge.

7.18 MASS SPECTROMETER OPERATING ON A LINEAR ACCELERATOR PRINCIPLE

A mass spectrometer in which the ions to be separated absorb maximum energy through the effect of alternating electric fields which are parallel to the path of the ions. These ions are then separated from other ions with different mass/charge by an additional electric field.

7.19   STATIC FIELDS MASS SPECTROMETER
An instrument which can separate a selected ion beam with fields which do not vary with time. The fields are generally both electric and magnetic.

7.20   DYNAMIC FIELD(S) MASS SPECTROMETER
A mass spectrometer in which the separation of a selected ion beam depends essentially on the use of fields, or a field, varying with time. These fields are generally electric.

7.21   PROLATE TROCHOIDAL MASS SPECTROMETER
A mass spectrometer in which the ions of different mass/charge are separated by means of crossed electric and magnetic fields in such a way that the selected ions follow a prolate trochoidal path.
Note: The usual term '*Cycloidal*' used sometimes is incorrect because the path used is not cycloidal. A cycloid is a special case of a trochoid.

7.22   CROSSED ELECTRIC AND MAGNETIC FIELDS
Electric and magnetic fields with the electric field at right angles to the magnetic field direction.

7.23   RESOLUTION: 10 PER CENT VALLEY DEFINITION
Let two peaks of equal height in a mass spectrum at masses m and $m - \Delta m$ (a.m.u.) be separated by a valley which at its lowest point is just 10 per cent of the height of either peak. For similar peaks at a mass exceeding m, let the height of the valley at its lowest point be more (by any amount) than 10 per cent of either peak height. Then the resolution (10 per cent valley definition) is $m/\Delta m$. It is usually a function of m. $m/\Delta m$ should be given for a number of values of m. This definition implies that, for an isolated symmetrical peak, at a distance $\pm \frac{1}{2}\Delta m$ along the mass scale from the peak maximum the peak height is 5 per cent of the maximum peak height.

7.24   RESOLVING POWER (MASS)
The ability to distinguish between ions differing in the quotient mass/charge by a small increment. It may be characterized by giving the peak width, measured in mass units, expressed as a function of mass, for at least two points on the peak, specifically at fifty per cent and for five per cent of the maximum peak height.

7.25   ELECTRON IMPACT IONIZATION
Ionization resulting from the interaction of an electron with any particle, e.g. a molecule or atom.

7.26   IONIZING VOLTAGE
The voltage difference through which electrons are accelerated before they are used to bring about electron impact ionization.
Note: To obtain the true ionizing voltage corrections for any contact or surface potentials must be made.
Note: Thw term '*electron energy*' is frequently used in place of 'ionizing voltage'.

7.27   FIELD IONIZATION
Ionization resulting from the effect of a very strong electric field on any particle. The strong electric field may produce ionization in space or in a region very close to a metal or

other surface.

7.28 PHOTOIONIZATION
Ionization resulting from the interaction of a photon with any particle which is, in consequence, ionized.

7.29 THERMAL IONIZATION
Ionization of particles brought about by a high temperature — for example, emission of ions from an adsorbed layer on an incandescent metal surface.

7.30 CHEMICAL IONIZATION
Ionization resulting from the collision of a particle with a positively or negatively charged ion.

7.31 CHEMI-IONIZATION
Ionization resulting from the collision of a particle with a neutral species (generally excited) such as a metastable helium atom.

7.32 SPARK SOURCE IONIZATION
Ionization resulting from a spark between electrodes.

7.33 IONIZATION BY SPUTTERING
Ionization by bombardment of a solid specimen with accelerated ions or electrons or fast neutrals.

7.34 LASER BEAM IONIZATION
Ionization by irradiation of a specimen with a laser beam.

7.35 FARADAY CUP (OR CYLINDER) COLLECTOR
A hollow collector, open at one end and closed at the other, used to collect beams of ions.

7.36 SECONDARY ELECTRON MULTIPLIER
A device to multiply current in an electron beam (or in a photon or particle beam by first conversion to electrons) by incidence of accelerated electrons upon the surface of an electrode which yields a number of secondary electrons greater than the number of incident electrons. These electrons are then accelerated to another electrode (or to another part of the same electrode) which in turn emits further secondary electrons so that the process can be repeated.

7.37 VIBRATING REED ELECTROMETER
A device to measure small currents which uses a vibrating reed forming part of a capacitor which has, in consequence, a periodically varying capacity allowing the measured signal to be modulated for a.c. amplification.

7.38 PHOTOGRAPHIC PLATE RECORDING
The recording of ion beams by allowing them to strike a photographic plate which is subsequently developed.

7.39 MASS SPECTRUM
This may refer to:
(i) The spectrum produced by a mass spectrometer which shows ion current as a function of the quotient mass/charge as a series of peaks corresponding to different ions.

(ii) The spectrum produced by a mass spectrograph which shows a series of lines on a photographic plate or film.

(In the limiting case a mass spectrum may only show a single ionic species rather than a series of lines or peaks, but this is unlikely in practice.)

Generally the term 'mass spectrum' refers to a spectrum of positive ions.

7.40 NEGATIVE-ION MASS SPECTRUM

A mass spectrum of negative ions.

7.41 MOLECULAR ION

The ion produced when a molecule gains or loses an electron.

7.42 MOLECULAR CATION

A molecular ion produced by the loss of one electron from a molecule.

7.43 MOLECULAR ANION

A molecular ion produced when a molecule gains an electron.

7.44 REARRANGED MOLECULAR ION

A molecular ion which has rearranged to a structure different from that of the original molecule.

7.45 PARENT ION

The ion precursor or progenitor of a fragment ion or of a metastable intermediate. The term is generally given the same meaning as 'molecular ion'.

7.46 PRECURSOR OR PROGENITOR ION

The precursor or progenitor of a fragment ion or of a metastable intermediate.

Note: A fragment ion may be the precursor or progenitor of other fragment ions.

7.47 FRAGMENT ION

An ion produced by the loss of one or more fragments from a parent molecular ion.

7.48 ISOTOPIC ION

In inorganic mass spectrometry this term generally means an ion containing one or more atoms of a less abundant isotope. In spark source mass spectrometry involving an element or elements with isotopes, any ion of any nuclide is an isotopic ion.

7.49 REARRANGEMENT ION

An ion with a structure not obtainable from the parent ion by the simple cleavage of bonds.

7.50 METASTABLE DECOMPOSITION

The decomposition of an ion of mass m_1 into an ion of mass m_2 ($m_2 < m_1$) occurring during the passage of the ion through the mass spectrometer. The decomposition is on a time-scale longer than that generating a normal fragment ion.

7.51 METASTABLE ION PEAK (OR METASTABLE PEAK)

The peak resulting from a metastable decomposition, often referring to the peak at $m^* = m_2^2/m_1$ resulting from decomposition in a single-focusing magnetic deflection mass spectrometer. It may also be observed in time-of-flight instruments by means of suitable retarding voltages.

7.52 APPEARANCE ENERGY (OR APPEARANCE POTENTIAL)

The lowest energy which must be imparted to the parent molecule to cause it to produce a particular specified ion. This energy, usually stated in electron volts, may be imparted by electron impact, by photon impact, or in other ways.

Note: It is recommended that the term '*appearance energy*' should replace the term '*appearance potential*' and that the energy should be stated in SI units.

7.53 BASE PEAK

The most intense peak in a mass spectrum. This term may be applied to the spectra of pure substances or mixtures.

7.54 INTENSITY RELATIVE TO BASE PEAK

The ratio of the ion current of a peak to that of the base peak. A process of normalization is generally used with the base peak current taken as 100.

7.55 PEAK HEIGHT

The height of a recorded peak in a mass spectrum.

7.56 TOTAL ION CURRENT

(a) *After mass analysis*

The sum of all the separate ion currents carried by the different ions contributing to the spectrum.

(b) *Before mass analysis*

The sum of all the separate ion currents for ions of the same sign (usually positive) before mass analysis.

7.57 ADDITIVITY OF MASS SPECTRA

The process by which each chemical species present in the ion source at a certain partial pressure makes a contribution to the total mass spectrum which is the same as that which it would give if that chemical species alone were present in the ion source at a pressure equal to this certain partial pressure.

7.58 INTERFERENCE

The modifying effect on the mass spectrum of a particular chemical species due to the presence of other chemical species in the ion source.

7.59 ION-MOLECULE REACTION

A chemical interaction between a positive or negative ion and an uncharged molecule.

7.60 AUTO-IONIZATION

The spontaneous ionization of an atom, molecule, or fragment of a molecule which is in a sufficiently excited state.

Note: The term 'Pre-ionization' is also used with the same meaning, particularly in the case of molecules.

7.61 IONIZATION EFFICIENCY CURVE

A curve showing the ion current of a particular ion plotted against the energy of the ionizing electrons or photons.

7.62 SENSITIVITY

(a) *Ion source*

The value obtained when the ion current (before any amplification) for a specified ion from a specified species is divided by the partial pressure of the species in the ion source.

(b) *Inlet system*

The sensitivity with respect to an inlet system for a gaseous sample should be given as a similar quotient using the partial pressure of the species in the inlet system or reservoir system.

(c) *Direct probe*

The sensitivity using a direct probe inlet should be given by stating the ion current for a specified ion from a specified species under specified conditions of operation.

It is usually very difficult to measure the partial pressure of the species in the ion source under standard operating conditions. For practical purposes it may be better to state the total ion current for a specified species using the direct probe inlet or other types of inlet systems.

7.63 BACKGROUND MASS SPECTRUM

The mass spectrum observed when no sample is intentionally introduced into the mass spectrometer or spectrograph.

7.64 α-CLEAVAGE

Fission adjacent to a heteroatom or functional group producing a radical and an ion.

7.65 β-CLEAVAGE

Fission next but one to a heteroatom or functional group producing a radical and an ion.

7.66 McLAFFERTY REARRANGEMENT

β-Cleavage with concomitant specific transfer of a γ-hydrogen atom in a six-membered transition state in mono-unsaturated systems irrespective of whether the arrangement is formulated by a radical or an ionic mechanism and irrespective of the position of the charge.

7.67 SYMBOLS USED IN MASS SPECTROMETRY

Symbol	Meaning
m/e	The quotient mass/charge
m^*	(i) Metastable peak (an asterisk denotes metastability)
	(ii) Value along the m/e scale for a metastable peak
*	Denotes a decomposition, the occurrence of which is supported by a metastable peak
$+\cdot$ or $\overset{+}{\cdot}$	Odd-electron ion
$+$	Even-electron ion
⌒	Single electron movement
⌢	Electron-pair movement
M	Mass number of molecular ion

8. RECOMMENDED NOMENCLATURE FOR TITRIMETRIC ANALYSIS*

8.01   ACIDIMETRY

The determination of a substance by titration with an acid.

8.02   ALKALIMETRY

The determination of a substance by titration with a base.

Comment. The term acidimetry has opposite meanings in different countries. For instance, in Britain and in the U.S.A. it is used in both senses, i.e. the determination *of* acid and determination *with* acid. In France acidimetry generally means determination *of* acid, and this appears to be the original meaning of the term. However, all other usages of similar terms imply titration *with*, e.g. argentimetry; accordingly, to maintain consistency it is recommended that acidimitry means titration *with*. The same remarks apply to alkalimetry.

It is recommended that terms ending with "-metry" should end in "-imetry" where possible, but it is appreciated that not every term can be standardized in this way.

8.03   AMINOPOLYCARBOXYLIC ACIDS

These are compounds containing the group

$$\begin{array}{c} HOOC-CH_2 \\ \diagdown \\ N- \\ \diagup \\ HOOC-CH_2 \end{array}$$

which comprises the most important class of chelate-forming titrants for the determination of metals. The most widely used compound of this type is *ethylenedinitrilotetraacetic acid*

$$\begin{array}{c} HOOC-CH_2 CH_2-COO^- \\ \diagdown \diagup \\ H-\overset{+}{N}-CH_2-CH_2-\overset{+}{N}-H \\ \diagup \diagdown \\ ^-OOC-CH_2 CH_2-COOH \end{array}$$

known as EDTA (and by various trade names). In practice the disodium salt of ethylene-dinitrilotetraaacetic acid (Na_2H_2Y) is used in place of the acid itself and it may also be designated EDTA.[†]

---

[*] Based on the approved Nomenclature published in *Pure and Applied Chemistry*, Vol. 18, No. 3 (1969), pp. 427 - 436.

[†] In accordance with established scientific practice the term *complexone* (plural *complexones*) is widely used to denote a complexing agent of the type of an aminopolycarboxylic acid. The word 'Complexone' is also registered as a trade name and reserved for certain commercial products marketed by the Uetikon Chemical Company, Switzerland.

8.04 BACK-TITRATION
The titration of an unreacted standard solution that has been added in excess to a sample.

8.05 BLANK TITRATION
A titration carried out on a solution identical with the sample solution (i.e. in volume, acidity, amount of indicator, etc.) except for the sample itself, which has been omitted.

8.06 BUFFER CAPACITY OR BUFFER INDEX
The capacity of a solution to resist changes in pH on addition of acid or base, which may be expressed numerically as the number of moles of strong acid or strong base required to change the pH by one unit when added to one litre of the specified buffer solution.

8.07 COMPARISON SOLUTION
(a) A solution having the same volume and indicator concentration as the solution being titrated, and of appropriate composition, which is used to detect the point at which the colour (or other property) of the solution being titrated begins to deviate from its initial colour (or other property), thus allowing the end-point to be found more precisely than otherwise.

(b) More specifically, a solution having the same composition as the solution being titrated at the equivalence-point, used to locate the equivalence-point as accurately as possible by matching some property of the two solutions; the entire composition of the comparison solution need not be the same as that of the solution being titrated (although this is desirable), the essential requirement being that the concentration of the substance determining the colour (or other property) of the indicator is the same in the comparison solution as in the solution being titrated at the equivalence-point.

8.08 COMPLEXIMETRY (COMPLEXOMETRY)
Titration with, or of, a substance capable of forming a soluble complex that is only slightly dissociated.

8.09 CONTROL TITRATION
A titration of a known amount of substance with a standard solution, made to determine the effect of variable factors and foreign substances on the accuracy of the titration.

8.10 DESIGNATED VOLUME
The designated volume is the volume at the particular temperature at which the volumetric glassware was calibrated; this temperature is usually taken as $20^\circ C$, but may be $25^\circ C$ or $27^\circ C$ (as in some tropical countries).

8.11 END-POINT
The point in a titration at which some property of the solution (as, for example, the colour imparted by an indicator) shows a pronounced change, corresponding more or less closely to the equivalence-point. The end-point may be represented by the intersection of two lines or curves in the graphical method of end-point determination (see *End-point detection*).

8.12 END-POINT DETECTION

8.12.01 AMPEROMETRIC END-POINT
The course of the reaction is monitored by means of a dropping mercury electrode, rotating platinum or other (concentration polarized) polarographic microelectrode as the indicator electrode in conjunction with a suitable reference electrode. The potential of the indicator electrode is set so as to register a diffusion current for the monitored ion and

a plot is made of diffusion current against the amount of titrant added. End-points are generally located by extrapolation at changes in the slope of the titration curve. *Polarimetric* is regarded as synonymous with *Amperometric* but it is not recommended because of possible confusion with terms used for the measurement of polarization (optical and electrochemical).

8.12.02 BI-AMPEROMETRIC END-POINT *

The course of the reaction is monitored by observing the current flowing between two similar (usually platinum) electrodes to which a small potential difference has been applied. The virtual removal of one component of a reversible redox couple from the system (or the appearance of a redox couple) at the end-point causes the current to cease abruptly (or to appear suddenly).

8.12.03 CHRONOPOTENTIOMETRIC END-POINT

The course of the reaction is monitored by means of a relatively large concentration-polarized electrode (usually Hg or Pt) in conjunction with a suitable reference electrode. A source of constant current is applied to the electrodes and the time required for the potential of the indicator electrode to transit from a predetermined value to a higher one is observed. A plot is made of the square root of transition time against titrant added. End-points are generally located by extrapolation at changes in slope in the titration curve.

8.12.04 CONDUCTIMETRIC END-POINT

The course of the reaction is monitored by measuring the conductance (reciprocal of the ohmic resistance) of the titration medium between two inert electrodes (usually platinized platinum) immersed in the reaction medium. A plot is made of conductance against titrant added. End-points are generally located by extrapolation at changes in slope in the titration curve.

8.12.05 FLUORIMETRIC (FLUOROMETRIC) END-POINT

The course of the reaction is monitored by means of changes in fluorescence of the reacting system either visually or photometrically while the solution is irradiated by a suitable activating source such as a mercury vapour lamp. Where instrumental detection is used, a plot is made of instrument response against titrant added. End-points are generally located by extrapolation at changes in slope in the titration curve.

8.12.06 HIGH-FREQUENCY END-POINT

The course of the reaction is monitored by a modification of the a.c. conductimetric technique in which the frequency of oscillation is in the megacycle per second (MHz) range. Changes in grid or anode current of the oscillating valve, or changes in frequency induced by chemical change, may be monitored and normally the electrodes (or tuned coil) are located outside the titration vessel. A plot is made of instrumental response against titrant added. End-points are generally located by extrapolation at changes in slope in the titration curve. (*Radio-frequency* is regarded as synonymous with *High-frequency*.)

8.12.07 NEPHELOMETRIC END-POINT

The course of the reaction in a precipitation system is monitored by measuring the light

---

* The term *bi-amperometric* is now recommended instead of *dead-stop*. For this and other terms in electrochemical analysis see the Classification and Nomenclature of Electroanalytical Techniques in Section 19.

scattered at right angles to the incident beam. A plot is made of instrument response against the amount of titrant added. End-points are generally located by extrapolation at changes in slope in the titration curve.

8.12.08 PHOTOMETRIC END-POINT

The course of the reaction is monitored by measuring the (optical) absorption of the titration medium within the selected narrow waveband. A plot is made of absorbance against titrant added. End-points are generally located by extrapolation of changes in slope in the titration curve. The monitor waveband is selected to match an absorption band of the titrant, titratable solute, reaction product, or added indicator.

8.12.09 POTENTIOMETRIC END-POINT

The course of the reaction is monitored by means of an indicator electrode (polarizable by one or more of the reacting ions) measured against a suitable reference electrode. The potential difference between the two electrodes is recorded. A plot is made of potential difference against titrant added. End-points are located at points of maximal slope on the titration curve.

8.12.10 RADIOMETRIC END-POINT

The course of the reaction is monitored radiochemically by adding a radiactive indicator which may be precipitated, or dissolved, at the equivalence-point, thus changing the radio-activity of the solution phase. A plot is made of instrumental response against titrant added. End-points are generally located by extrapolation at changes in slope in the titration curve. (In some instances the titratable solute or the titrant may be radioactive, in which case no added indicator may be required.)

8.12.11 THERMOMETRIC END-POINT

The course of the reaction is monitored by means of a sensitive temperature measuring device (thermistor, thermocouple or thermometer) immersed in the reaction medium in a thermally isolated vessel. A plot is made of the response of the monitor device against titrant added. End-points are generally located by extrapolation at changes in slope in the titration curve. (*Enthalpimetric* is regarded as synonymous with *Thermometric*.)

8.12.12 TURBIDIMETRIC END-POINT

This is a technique similar to nephelometric titration in that it involves a precipitation system, but the course of the reaction is monitored by measurements of transmitted rather than scattered radiation.

8.12.13 VISUAL END-POINT

The course of the reaction is monitored by visual observation of the colour (or other) change of an added indicator on neutralization, oxidation-reduction, precipitation or complexation. (In some instances the titratable solute or the titrant may be sufficiently coloured not to require the addition of an indicator.)

8.13 EQUIVALENCE-POINT

The point in a titration at which the amount of titrant added is chemically equivalent to the amount of substance titrated. (*Stoichiometric (stoicheiometric) end-point* and *Theoretical end-point* are synonymous with *Equivalence point*.)

8.14 FACTOR WEIGHT

A weight of sample such that the titration volume in millilitres represents the percentage of constituent or a simple multiple or fraction of the percentage.

8.15 FORMALITY

The number of gram formula weights of the reacting substance in one litre of solution, the formula being specified whenever there is a possibility of ambiguity.

8.16 INDICATOR (VISUAL)

A substance which indicates a visual change at or near the equivalence point of a titration and which is, ideally, present in sufficiently small concentration not to consume an appreciable amount of the titrant in passing through its transition range.

8.17 INDICATORS (VISUAL), TYPES OF

8.17.01 ONE-COLOUR INDICATOR

An indicator which exhibits a colour on only one side of its transition interval and is colourless on the other, or which exhibits a deeper or a less intense shade of the same colour on one side of the interval.

8.17.02 TWO-COLOUR INDICATOR

An indicator exhibiting two different colours, one on each side of the transition interval.

8.17.03 ACID-BASE INDICATOR

An indicator which is itself an acid or base and which exhibits a colour change on neutralization by a base or acid at or near the equivalence-point of a titration.

8.17.04 ADSORPTION INDICATOR

An indicator which is adsorbed or desorbed, with concomitant colour change, by a precipitation system at or near the equivalence-point of a titration.

8.17.05 CHEMILUMINESCENT INDICATOR

An indicator (acid-base or other type) exhibiting chemiluminescence, or a quenching of chemiluminescence, at or near the equivalence-point.

8.17.06 EXTRACTION INDICATOR

An indicator (acid-base or other type) which is abruptly extracted from one liquid phase into another at or near the equivalence-point of a titration. The indicator need show no colour change in the process. In some instances a titrant may serve as its own indicator.

8.17.07 FLUORESCENT INDICATOR

An indicator (acid-base or other type) which, while being activated by radiation of a suitable wavelength, exhibits a change in fluorescence emission at or near the equivalence-point of a titration.

8.17.08 METALLOCHROMIC INDICATOR

An indicator which is itself a complexing agent and which exhibits a colour change when it reacts with metal ions or has them removed from its complex with them at or near the equivalence-point of a complexometric or precipitation titration.

8.17.09 METALLOFLUORESCENT INDICATOR

A special type of metallochromic indicator which is itself a complexing agent and which, whilst undergoing suitable irradiation, exhibits a change in its fluorescence emission when it reacts with metal ions or has them removed from its complex with them at or near the equivalence-point of a complexometric or precipitation titration.

8.17.10 MIXED INDICATOR

A mixture of indicators of the same function chosen so that their transition intervals are approximately coincident and of such a composition that the resultant overall colour change of the mixture is more easily distinguished than that of either indicator used separately.

8.17.11 OXIDATION-REDUCTION (REDOX) INDICATOR

An indicator which is capable of being oxidised or reduced and which undergoes a colour change in the process at or near the equivalence-point.

8.17.12 PRECIPITATION INDICATOR

An indicator which precipitates from solution in a readily visible form at or near the equivalence-point.

8.17.13 SCREENED INDICATOR

A mixture of an indicator (acid-base or other type) and a suitable indifferent dyestuff chosen so as to screen out unwanted parts of the visible range spectrum transmitted by the indicator in one of its forms.

8.18 INDICATOR BLANK OR INDICATOR CORRECTION

The amount of titrant (usually in terms of volume) required to produce the same change in the indicator as that taken to mark the end-point in the titration of the sample under the same conditions. It is not necessarily the same as the total blank (covering all the steps of an analysis), which may include the effect of other factors, such as the presence of small amounts of reacting substances in the water (or other solvent) or reagents.

8.19 INDICATOR, RADIOACTIVE

A radioactive substance which functions as an adsorption, precipitation, or extraction indicator.

8.20 LEVEL OF TITRATION

The order (10^{-x}) of concentration (normality, formality) at which the solution of the titrant is used, e.g. 10^{-1}, 10^{-2}, or 10^{-3}....

8.21 MASKING AGENT

A substance preventing the interfering reaction of one or more foreign substances in a determination by conversion into soluble complexes, different oxidation states, or other unreactive forms.

8.22 STANDARDIZATION

The process of finding the concentration of an active agent in solution, or the reacting strength of a solution in terms of some substance, usually by titration of a known amount of this substance which is pure or has a known reaction value.

8.23 STANDARD SOLUTION

A solution having an accurately known concentration, or an accurately known titre.

8.23.01 PRIMARY STANDARD SOLUTION

A standard solution, prepared from a primary standard substance, whose concentration is known from the weight of that substance in a known volume (or mass) of the solution.

8.23.02 SECONDARY STANDARD SOLUTION

A solution whose concentration or titre has been obtained by standardization, or which has been prepared from a known mass of a secondary standard substance.

8.24 STANDARD SUBSTANCE

8.24.01 PRIMARY STANDARD

A substance of high purity which, by stoicheiometric reaction, is used to establish the reacting strength of a titrant, or which itself can be used to prepare a titrant solution of accurately known concentration.

8.24.02 SECONDARY STANDARD

A substance used for standardization, whose content of the active agent has been found by comparison against a primary standard.

8.25 TITRANT

The solution containing the active agent with which a titration is made.

8.26 TITRATION

The process of determining a substance A by adding increments of substance B (almost always as a standardized solution) with provision for some means of recognizing the point at which all of A has reacted, thus allowing the amount of A to be found from the known amount of B added up to this point, the reacting ratio of A to B being known from stoicheiometry or otherwise. The reverse process — incremental addition of A to B — is seldom applied, except in standardization titrations.

8.27 TYPES OF TITRATIONS

8.27.01 ACID-BASE TITRATIONS

Titrations involving the transfer of protons (Brønsted-Lowry) or electron-pairs (Lewis) from one reacting species to the other in solution.

8.27.02 ACIDIMETRIC TITRATION

An acid-base titration in which a base is titrated *with* a standard solution of an acid.

8.27.03 ALKALIMETRIC TITRATION

An acid-base titration in which an acid is titrated *with* a standard solution of an alkali.

8.27.04 CHELATOMETRIC TITRATION

A titration in which a soluble chelate complex is formed; this is a special case of complexometric titration.

8.27.05 COMPLEXIMETRIC (COMPLEXOMETRIC) TITRATION

A titration involving the formation of a soluble complex between a metal ion and a complexing agent.

8.27.06 COULOMETRIC TITRATION

A titration technique in which the titrating agent (acid-base or other type) is generated electrolytically *in situ* or externally and is not added as a standard solution as in the conventional manner. Time and current measurements are generally made in place of volume

8.27.07 INDIRECT TITRATION

A titration (acid-base or other type) in which the entity being determined does not react directly with the titrant, but indirectly *via* the intermediacy of a stoicheiometric reaction with another titratable entity.

8.27.08 IODIMETRIC TITRATION

Titration with, or of, iodine (usually I_3^-). (Some authors restrict *iodimetry* to titration *with* a standard solution of iodine, and *iodometry* to titration *of* iodine; such restriction is not recommended.)

8.27.09 NON-AQUEOUS TITRATION

A titration (acid-base or other type) in which the solvent medium is one other than water and in which the concentration of the latter is minimal (say less than 0.5 per cent).

8.27.10 OXIDATION-REDUCTION (REDOX) TITRATION

A titration involving the transfer of one or more electrons from a donor ion or molecule (the reductant) to an acceptor (the oxidant).

8.27.11 PHASE TITRATION

A titration in which the entity being titrated is present in a two-phase (liquid) system and which is caused to become a single phase at or near the equivalence-point, or one in which a monophase containing two miscible components is caused to separate into a two-phase system by addition of a third component.

8.27.12 PRECIPITATION TITRATION

A titration in which the entity being titrated is precipitated from solution by reaction with the titrant.

8.27.13 WEIGHT TITRATION

A titration in which the amount of titrant is found from the weight of a standard solution required to reach the end-point.

8.28 TITRATION ERROR

The difference in the amount of titrant, or the corresponding difference in the amount of substance being titrated, represented by the expression:

$$\text{(End-point value - Equivalence-point value)}$$

8.29 TITRE (TITER)

The reacting strength of a standard solution, usually expressed as the weight (mass) of titrated substance equivalent to 1 cm^3 of the standard solution. (The term titre must not be used to denote the volume of the titrant consumed in a particular titration.)

8.30 TITRIMETRIC ANALYSIS

Analysis based on titration. Sometimes called *volumetric analysis* from the common method of measuring the titrant. This latter term is not recommended.

8.31 TITRIMETRIC CONVERSION FACTOR

The factor giving the amount (usually weight) of the titrated substance corresponding to a unit amount (usually cm^3) of the standard solution. This factor may be the formality, etc. of the standard solution multiplied by the milliequivalent value of the substance titrated or the titre of the standard solution.

8.32 TRANSITION INTERVAL

The range in concentration of hydrogen ion, metal ion, or other species over which the eye is able to perceive a variation in hue, colour intensity, fluorescence, or other property of a visual indicator arising from the varying ratio of the two relevant forms of the indicator. The range is usually expressed in terms of the negative decadic logarithm of the concentration (e.g. pH). For an oxidation-reduction indicator the transition interval is the corresponding range in oxidation-reduction potential.

9. REPORT ON THE STANDARDIZATION OF pH AND RELATED TECHNOLOGY*

9.1 INTRODUCTION

The standardization of pH has been achieved in Great Britain[1], in the United States[2-5], and recently also in Japan[6]. In their essential elements, the three standards are in agreement; only differences of detail exist. It is the object of this report to summarize, for the information of all national groups interested in pH, the extent of standardization already achieved and to indicate how the area of agreement now reached could perhaps be extended to other conceptions related to pH.

9.2 SYMBOLS

There already exists international agreement that pH should be written and printed on line in roman type. We recommend that, with the unique exception of pH, the operator p (printed in roman) should denote $-\log_{10}$. For example:

9.2.01 pm_H means $-\log_{10} m_H$

9.2.02 $pm_H \gamma_{H,Cl}$ means $-\log_{10}(m_H \gamma_{H,Cl})$

9.2.03 $pm_H \gamma^2_{H,Cl}$ means $-\log_{10}(m_H \gamma^2_{H,Cl})$

where m denotes molality and γ denotes mean activity coefficient (molality scale).

9.3 OPERATIONAL DEFINITION OF pH

In all existing national standards the definition of pH is an operational one. The electromotive force E_X of the cell

     Pt,H$_2$ | solution X | concentrated KCl solution | reference electrode

is measured and likewise the electromotive force E_S of the cell

     Pt,H$_2$ | solution S | concentrated KCl solution | reference electrode

both cells being at the same temperature throughout and the reference electrodes and bridge solutions being identical in the two cells. The pH of the solution X, here denoted by pH(X), is then related to the pH of the solution, here denoted by pH(S), by the definition

$$\mathrm{pH}(X) = \mathrm{pH}(S) + \frac{(E_X - E_S)F}{RT \ln 10} \qquad (9.3.1)$$

where R denotes the gas constant, T the absolute temperature, and F the faraday. In this equation the numerator and denominator must be expressed in the same units so that the pH

---

* Based on the approved Report published in *Pure and Applied Chemistry*, Vol.1, No.1 (1960), pp. 163 - 168.

difference here defined is a pure number.

The hydrogen electrode in both cells may be replaced by identical reversible hydrogen-ion responsive electrodes, e.g., glass or quinhydrone. The two salt bridges may be of any molality not less than about 3.5 mol kg^{-1}, provided they are the same.

9.4 STANDARDS FOR MEASUREMENT OF pH

The difference between the pH of two solutions having been defined as above, the definition of pH can be completed by assigning a value of pH at each temperature to one chosen solution called the primary standard. In the British and Japanese Standards this is the definition of pH, the primary standard being a one-twentieth molar solution of pure potassium hydrogen phthalate. The pH of this solution is defined as having the value 4 exactly at 15°C. At any other temperature t°C between 0°C and 60°C its pH is defined by

$$\text{pH} = 4.00 + \frac{1}{2}\left(\frac{t-15}{100}\right)^2 \qquad 9.4.01$$

This formula is for all practical purposes equivalent to one given elsewhere[7]. The question how this formula might be extended to temperatures higher than 60°C has been raised but not yet formally answered. The American Standard described below is in practically complete agreement with

$$0 < t < 55 \qquad \text{pH} = 4.00 + \frac{1}{2}\left(\frac{t-15}{100}\right)^2 \qquad 9.4.02$$

$$55 < t < 95 \qquad \text{pH} = 4.00 + \frac{1}{2}\left(\frac{t-15}{100}\right)^2 - \frac{t-55}{500} \qquad 9.4.03$$

The American Standard specifies four standard solutions with pH values specified at each temperature covering the range of pH from 2 to 12. These are given in Table 9.4.

There is a proposal to add as secondary standards 0.05 molar potassium tetroxalate (pH ≈ 1.7) and saturated calcium hydroxide (pH ≈ 12).

If the definition of pH in section 9.3 is adhered to strictly, then the pH of a solution might be slightly dependent on which standard solution was used. In fact any such variation is too small to be of practical significance. Moreover the American acceptance of several standards allows the use of the following alternative definition of pH. The electromotive force of the cell

$$\text{Pt,H}_2 \mid \text{solution X} \mid \text{concentrated KCl solution} \mid \text{reference electrode}$$

is measured, and likewise the electromotive forces E_1 and E_2 of two similar cells with the solution X replaced by the standard solutions S_1 and S_2 such that the values of E_1 and E_2 are on either side of, and as near as possible to, E_x. The pH of solution X is obtained by assuming linearity between pH and E, that is to say

$$\frac{\text{pH}(X) - \text{pH}(S_1)}{\text{pH}(S_2) - \text{pH}(S_1)} = \frac{E_s - E_1}{E_2 - E_1}$$

This procedure is in fact useful and especially recommended when the hydrogen-ion-responsive electrode is a glass electrode.

TABLE 9.4*

The pH of standard buffers at temperatures from 0 - 95°C

t /°C	KH tartrate (satd. at 25°C)	0.05 KH phthalate	0.025 KH_2PO_4 + 0.025 Na_2HPO_4	0.01 Borax
0	–	4.01	6.98	9.46
5	–	4.01	6.95	9.39
10	–	4.00	6.92	9.33
15	–	4.00	6.90	9.27
20	–	4.00	6.88	9.22
25	3.56	4.01	6.86	9.18
30	3.55	4.01	6.85	9.14
35	3.55	4.02	6.84	9.10
40	3.54	4.03	6.84	9.07
45	3.55	4.04	6.83	9.04
50	3.55	4.06	6.83	9.01
55	3.56	4.07	6.84	8.99
60	3.56	4.09	6.84	8.96
70	3.58	4.12	6.85	8.93
80	3.61	4.16	6.86	8.89
90	3.65	4.20	6.88	8.85
95	3.68	4.23	6.89	8.83

* All concentrations are quoted in mol dm^{-3} of solution. In the IUPAC *Manual of Symbols and Terminology for Physicochemical Quantities and Units* (Butterworths, 1973) values of pH are listed to 4 significant figures for five standard buffer solutions for which the concentrations are expressed in mol kg^{-1}.

9.5 IONIC ACTIVITY COEFFICIENTS

As a prelude to the interpretation of pH it is expedient to say something about ionic activity coefficients. For the sake of brevity we shall restrict our detailed discussion to solutions containing only 1-1 electrolytes. The extension to the general case would require more complicated formulae without, however, requiring any new physical principles.

Electromotive force measurements on cells without liquid-liquid junctions (or other less direct experimental techniques) lead to experimental values of mean activity coefficients of an electrolyte but not to ionic activity coefficients. In particular the electromotive force of the cells

$$H_2,Pt \mid \text{solution containing } H^+ \text{ and } Cl^- \mid AgCl \mid Ag$$

leads directly to the value of $m_H m_{Cl} \gamma^2_{H,Cl}$ and, if the value of m_{Cl} is known, indirectly to that of $m_H \gamma^2_{H,Cl}$.

It is permissible, and often convenient, to write conventionally

$$\gamma_{H,Cl}^2 \quad \gamma_H\gamma_{Cl} \qquad \qquad 9.5.01$$

$$\gamma_{Na,Cl}^2 \quad \gamma_{Na}\gamma_{Cl} \qquad \qquad 9.5.02$$

$$\gamma_{K,Cl}^2 \quad \gamma_K\gamma_{Cl} \qquad \qquad 9.5.03$$

and so on, but the values of the ionic activity coefficients γ_H, γ_{Na}, γ_K, and γ_{Cl} are not uniquely defined. In each solution an arbitrary value may be assigned to the γ of any one chosen ionic species. This does not mean that one convention may not be more convenient than another. In particular all 1-1 electrolytes present in a solution of total molality less than 0.01 have, to an accuracy sufficient for most purposes, equal mean activity coefficients. Under these conditions it is natural to equate the ionic activity coefficients to the common mean activity coefficient; in fact any other convention would be far-fetched.

At higher molalities different electrolytes in the same solution have different mean activity coefficients and consequently the convention must be less simple.[9] If we use M to denote a cation and X an anion, then

$$\frac{\gamma_H}{\gamma_M} \equiv \frac{\gamma_H\gamma_X}{\gamma_M\gamma_X} \equiv \frac{\gamma_{H,X}^2}{\gamma_{M,X}^2} \qquad 9.5.04$$

is thermodynamically well defined as is, of course, $\gamma_M\gamma_X \equiv \gamma_{M,X}^2$. The convention for defining γ_H is

$$\ln \gamma_H = \frac{\Sigma_M m_M \ln(\gamma_H/\gamma_M) + \Sigma_X m_X \ln(\gamma_H\gamma_X)}{\Sigma_M m_M + \Sigma_X m_X} \qquad 9.5.05$$

with the analogous definition for a typical anion, say Br^-:

$$\ln \gamma_{Br} = \frac{\Sigma_M m_M \ln(\gamma_M\gamma_{Br}) + \Sigma_X m_X \ln(\gamma_{Br}/\gamma_X)}{\Sigma_M m_M + \Sigma_X m_X} \qquad 9.5.06$$

These conventional relationships automatically satisfy the thermodynamic requirement

$$\ln \gamma_H + \ln \gamma_{Br} = 2 \ln \gamma_{H,Br} \qquad 9.5.07$$

When there is only a single electrolyte present, say HBr, these conventions reduce to

$$\ln \gamma_H = \ln \gamma_{Br} = \ln \gamma_{H,Br} \qquad 9.5.08$$

This convention is the simplest possible which treats all ionic species on a par. It is aesthetically satisfying, but it has little practical utility because the values of all the mean activity coefficients of the several electrolytes present are usually not known. (In the more general case of ions with various charges the formulae are more complicated.)

We are therefore driven to using a much simpler, though less symmetrical, convention. One that has proved especially useful is to define the activity coefficient of the *chloride* ion, at an ionic strength I not exceeding 0.1, by

$$\log \gamma_{Cl} = -\frac{AI^{\frac{1}{2}}}{1 + \rho I^{\frac{1}{2}}} \qquad 9.5.09$$

where A has at each temperature the value given by the theory of Debye and Hückel and ρ has a specified value; that recommended is $\rho = 1.5 \text{ mol}^{-\frac{1}{2}} \text{ kg}^{-\frac{1}{2}}$.

Once the value of the activity coefficient of the chloride ion has been conventionally defined, this value may be combined with the value of $m_H \gamma^2_{H,Cl} = m_H \gamma_H \gamma_{Cl}$ to obtain the value of $m_H \gamma_H$ or of $pm_H \gamma_H = -\log_{10} m_H \gamma_H$.

9.6 pH OF STANDARD SOLUTIONS

We have stated that pH values have been assigned to four standard solutions in the American Standard and to one of these in the British and Japanese standards. We have as yet said nothing of how these values were chosen. They are in fact defined for each standard solution S by

$$\text{pH}(S) = pm_H \gamma_H \qquad 9.6.01$$

where the quantity on the right has been determined according to the convention of 9.5.

9.7 APPROXIMATE INTERPRETATION OF pH

As a consequence of the relationship defining the pH of standard solutions, viz.

$$\text{pH} = pm_H \gamma_H \qquad 9.7.01$$

it can be verified that this same relationship holds with an accuracy of ± 0.02 or better for all aqueous solutions of total ionic strength not exceeding 0.1, provided that the pH lies between 2 and 12. Outside this pH range a more complicated recipe would be needed in any attempt to correlate pH with $pm_H \gamma_H$.

9.8 THE ABBREVIATION $p\alpha_H$

Some people may wish to abbreviate $pm_H \gamma_H$ (as conventionally defined in Section 9.5) to $p\alpha_H$. Other people have strong objections to this notation because it may revive controversies of the days before the nature of pH was properly understood.

REFERENCES

[1] *Brit. Standard*, 1647 (1950)
[2] *Chem. Rev.*, **48**, 1 (1948)
[3] *Compt. rend. 15e Conf. I.U.P.A.C.*, 118 (1949)
[4] *Compt. rend. 16e Conf. I.U.P.A.C.*, 72 (1951)
[5] Amer. Soc. Testing Materials, Method E70 - 52T (1952)
[6] *Japanese Industrial Standard*, Z8802 (1958)
[7] *J. Research Nat. Bur. Standards*, **36**, 47 (1946)
[8] *J. Research Nat. Bur. Standards*, **59**, 261 (1957)
[9] *J. Phys. Chem.*, **34**, 1758 (1930)

10. PRACTICAL MEASUREMENT OF pH IN AMPHIPROTIC AND MIXED SOLVENTS*

10.1 INTRODUCTION

Procedures analogous to those on which a practical pH scale has been based can be used profitably to establish operational acidity scales in certain other amphiprotic and mixed solvent media. A universal pH scale relating proton activity uniformly to the aqueous standard reference state is not yet a practical possibility, but separate scales for each medium can be achieved and will fulfil most of the requirements. The best choice of unit appears to be $p\alpha_H^*$ or $-\log(m_H \cdot {}_s\gamma_H)$, where ${}_s\gamma_H$ is referred to the standard state in each particular medium s. Data for reference solutions in 50 wt. per cent methanol and deuterium oxide are given below in Table 10.8.

10.2 THE OPERATIONAL pH SCALE

The well-recognized difficulty in reconciling a fundamental definition of the pH value with the practical experimental procedures for the routine measurement of acidity has led to the widespread adoption of an operational definition of the pH[1-3], viz.

$$pH(X) - pH(S) = \frac{(E_X - E_S)F}{RT \ln 10} \qquad 10.2.1$$

In this equation, X designates the solution of unknown pH and S the standard reference solution of known or assigned pH, while E is the electromotive force of a suitable pH cell consisting of an electrode reversible to hydrogen ions (usually a glass electrode, a hydrogen gas electrode, or quinhydrone electrode) coupled with a suitable reference electrode (commonly calomel-mercury or silver-silver chloride electrode). A bridge composed of a concentrated solution of potassium chloride usually connects the reference electrode with solution X or S when the cell is filled. The symbols R, T, and F represent the molar gas constant, the absolute temperature, and the faraday respectively.

10.3 INTERPRETATION OF THE MEASURED pH

Under optimum conditions, the potential difference across the diffusion junctions Solution X | conc. KCL solution and Solution S | conc. KCl solution can be considered to be equal, and then the difference of emf, $E_X - E_S$ is a useful formal measure of a relative hydrogen ion activity in the two solutions[4]; in other words, the left-hand side of equation 10.2.1 can then be written as $p\alpha_H(X) - p\alpha_H(S)$. To assure close equality of the junction potentials, solutions X and S should match closely in pH, ionic strength,

---

*Based on the approved report published in *Pure and Applied Chemistry*, Vol.18, No.3, pp. 421 - 425.

and composition, and neither should interact chemically with K^+ or Cl^- ions. In particular the pH should neither be greater than 11.5 nor less than 2.5, and the concentrations of non-electrolytes should be low and equal. Values of pH(S) for selected reference solutions correspond to a conventional hydrogen ion activity referred to the standard state for aqueous solutions:

$$pH(S) \equiv p\alpha_H(S) \qquad 10.3.1$$

This approach to pH standardization has received the endorsement of two Commissions of the International Union of Pure and Applied Chemistry[5]. Embracing the operational definition set forth in equation 10.2.1 it permits the experimental evaluation of pH(X) in a wide variety of media. Many if not most of these media differ so profoundly in composition from the standard reference solutions that the "relative hydrogen ion activity" obtained from pH(X) - pH(S) is virtually meaningless. For example, reproducible pH(X) values in non-aqueous media are often obtainable, but it is not possible to interpret these values usefully in terms of the proton level in the solution. It is the purpose of this report to recommend a procedure which, with further development, gives promise of permitting a useful interpretation, under optimum conditions, of practical pH numbers for many amphiprotic and mixed solvents. This approach, which has already been set forth elsewhere[4-6], has its roots in earlier proposals for the useful measurement of acidity in various media[7-9].

10.4 EXTENSION TO OTHER SOLVENTS

In general, the solvent should be amphiprotic, that is, be capable, like water, of either combining with protons released from acids or of furnishing protons to bases added to the medium. Furthermore, the experimental aspects of the method require that the hydrogen gas electrode and the silver-silver chloride electrode be thermodynamically reversible and stable in the medium. For the pH meter to be useful, the response of the glass electrode to hydrogen ions in the medium must also be affirmed. Solvents that meet these criteria include alcohols, alcohol-water mixtures, and deuterium oxide.

In order to establish a useful practical scale, it must be shown first that the liquid-junction potential may be reasonably constant and unaffected by a considerable variation in the acidity of the solution, as long as the solvent composition remains unchanged. A means of demonstrating this constancy has already been devised. It is then possible to determine relative hydrogen ion activities in the medium with the hydrogen electrode (cf. equation 10.2.1). It only remains to select a reference value which is consistent with the thermodynamics of the pH cell used and which will endow the measured pH(X) with a clear meaning in terms of chemical equilibria. An analogous procedure will serve for the establishment of a pD scale in deuterium oxide (D_2O).

The data needed are (i) the emf of the pH cell with liquid junction

$$Pt; H_2(g), \text{ solution X} \mid \text{concentrated KCl, calomel electrode} \qquad 10.4.1$$

where solution X is a buffer containing chloride ions in solvent s, (ii) the emf of the cell without a liquid junction

$$Pt; H_2(g), \text{ solution X, AgCl;Ag} \qquad 10.4.2$$

and (iii) the standard emf of cell 10.4.1 both in water ($_wE^O$) and in solvent ($_sE^O$). It should be noted that a knowledge of the difference of liquid-junction potentials (or E_j, when expressed in pH units) is sufficient to make equation 10.2.2 exact. In other words, if pH(S) were an exact (but conventional) $p\alpha_H$ in the aqueous medium, one could write

$$p\alpha_H(X) = pH(X) - E_j \qquad 10.4.3$$

where the "experimental" α_H would also refer to the aqueous standard state. Hereafter the designation (X) will be omitted.

It is further noted that measurements of cell 10.4.2 can yield two acidity functions, namely $p_w(\alpha_H\gamma_{Cl})$ and $p_s(\alpha_H\gamma_{Cl})$ depending on whether $_wE^o$ (aqueous standard state) or $_sE^o$ (standard state in solvent s) is chosen for the calculation. Similarly, molal activity coefficients are designated either $_w\gamma_i$ or $_s\gamma_i$, depending on the standard state used. The difference between the two acidity functions is always

$$p_s(\alpha_H\gamma_{Cl}) - p_w(\alpha_H\gamma_{Cl}) = 2 \log\, _m\gamma_{HCl} \qquad 10.4.4$$

where $_m\gamma_{HCl}$ is the "medium effect" for the transfer of hydrochloric acid from the aqueous standard state to the standard state in solvent s. It has been shown that measurements of pH, $p_w(\alpha_H\gamma_{Cl})$, and $_m\gamma_{HCl}$, together with a Debye-Hückel correction for interionic effects (small for dilute buffer solutions), can lead to the useful quantity δ, where

$$\delta \equiv E_j - \log\, _m\gamma_H \qquad 10.4.5$$

This quantity can also be derived from a comparison of "true" dissociation constants with "apparent" constants based on pH measurements[10]. Medium effects are characteristic only of the properties of substances in their standard states; hence, constancy of δ when the acidity varies at a fixed solvent composition is sufficient evidence to confirm the constancy of the liquid-junction potential.[6]

10.5 SELECTION OF A pH UNIT FOR AMPHIPROTIC SOLVENTS

Although some confidence in the values of δ is justifiable, there is no way of determining E_j and $_m\gamma_H$ individually. Hence, a $p\alpha_H$ referred always to the aqueous standard state regardless of the solvent, though eminently desirable, does not appear possible. Considerable effort and ingenuity have been expended on various means of evaluating the medium effect for hydrogen ions[11-13], but these methods have not yet reached such a degree of refinement as to warrant the establishment of a single general scale of pH. Furthermore, a barrier similar to that encountered with aqueous solutions precludes the experimental determination of hydrogen ion concentrations or pm_H.[4]

A quantity $p(m_H \cdot _s\gamma_H)$ or $p\alpha_H^*$ can, however, be derived formally from the experimental pH obtained from aqueous reference solutions, providing δ is known, by simple difference:

$$p\alpha_H^* = pH^* - \delta \qquad 10.5.1$$

The $p\alpha_H^*$ itself has a clear meaning and is consistent with equilibrium expressions involving (as is customary) the dissociation constant $p(_sK)$.

10.6 OPERATIONAL pH SCALE FOR AMPHIPROTIC SOLVENTS

The use of tabulated δ corrections to convert measured operational pH values into approximate $p\alpha_H^*$ values thus appears to be a practical possibility. Nevertheless a better procedure would involve the use of reference solutions prepared in a medium of the same solvent composition as the test solutions. In this way, errors attendant on the transfer of the glass electrode from the aqueous medium used for standardization to the nonaqueous test medium could be avoided. The establishment of an operational scale in each medium is thus envisioned.

$$pH^*(X) = pH^*(S) + \frac{(E_X - E_S)F}{RT \log 10} \qquad 10.6.1$$

10.7 REFERENCE SOLUTIONS FOR VALUES OF pH* AMPHIPROTIC SOLVENTS

As in water, the pH*(S) of each reference solution would be identified with $p\alpha_H^*$ determined independently for that solution. The latter is related formally to $p_s(\alpha_H \gamma_{Cl})$ by

$$p\alpha_H^* = p_s(\alpha_H \gamma_{Cl}) + \log {}_s\gamma_{Cl} \qquad 10.7.1$$

The activity coefficient ${}_s\gamma_{Cl}$ is predominantly a reflection of interionic forces, becoming unity at infinite dilution in the solvent s. In the absence of any physical means of evaluating single ion activities, the numerical value of ${}_s\gamma_{Cl}$ must depend upon some convention or formula accepted for this purpose.

The formula used in aqueous solutions[5] makes γ_{Cl} in solutions of ionic strength (I) less than 0.1 closely equal to the mean molal activity coefficient of sodium chloride at a molality of I. Furthermore, it is equivalent to assigning a value of 4.56 Å to the ion-size parameter a in the Debye-Hückel expression:

$$-\log \gamma_{Cl} = \frac{A'I^{\frac{1}{2}}}{1 + B'aI^{\frac{1}{2}}} \qquad 10.7.2$$

To preserve consistency, a convention for ${}_s\gamma_{Cl}$ in other amphiprotic solvents of sufficiently high dielectric constant to avoid ion-pair formation might appropriately utilize equation 10.7.2 with the same value of a selected for aqueous solutions. The values of the Debye-Hückel constants A' and B' must, of course, be modified as required by changes in the absolute temperature (T) and the dielectric constant (ε) and the density (d_o) of the solvent concerned. Their values are given by

$$A' = \frac{1.8246 \times 10^6}{(\varepsilon T)^{\frac{1}{2}}} d_o^{\frac{1}{2}} \qquad 10.7.3$$

and

$$B' = \frac{50.29}{(\varepsilon T)^{\frac{1}{2}}} d_o^{\frac{1}{2}} \qquad 10.7.4$$

10.8 ACCURACY OF PRACTICAL SCALES

By this procedure, conventional pa_H^* values for three reference solutions in 50 wt. per cent methanol-water have been determined from 10 to $40°C^{14}$. Likewise, pa_D values for two series of reference solutions in deuterium oxide (D_2O) have been derived[15]. The reference values, some of which are summarised in Table 10.8, can be identified with $pH^*(S)$ in Equation 10.6.1 or with $pD(S)$ in an analogous expression for pD in the solvent deuterium oxide.

TABLE 10.8 Conventional pa_H^* and pa_D values for reference solutions at 10, 25, and 40°C (m = molality)

Solution	pa_H^* and pa_D values at		
	10°C	25°C	40°C
In 50 wt. per cent methanol (pa_H^*)			
CH_3COOH (0.05m) CH_3COONa (0.05m), NaCl (0.05m)	5.52	5.49	5.50
Sodium hydrogen succinate (0.05m) NaCl (0.05m)	5.72	5.67	5.65
KH_2PO_4 (0.02m), Na_2HPO_4 (0.02m) NaCl (0.02m)	7.94	7.88	7.86
In deuterium oxide (pa_D)			
CH_3COOD (0.05m), CH_3COONa (0.05m)	5.25	5.23	5.23
KD_2PO_4 (0.025m), Na_2DPO_4 (0.025m)	7.50	7.43	7.38

The errors in $_sE°$, in the preparation of the solutions and materials, and in the experimental measurement of e.m.f endow $p_s(a_H\gamma_{Cl})$ with an aggregate uncertainty of about 0.004 pH unit at 25°C. The uncertainty of pH* or pD obtained from routine cell measurements by Eq. 10.6.1 will be considerably greater than this amount, however, in view of the unavoidable differences among the liquid-junction potentials when individual reference solutions and unknowns of different ionic composition are used. It is therefore unwise to expect an accuracy greater than 0.02 unit except under the most carefully controlled conditions.

REFERENCES

1. Am. Soc. Testing Materials, Method E70-52T (1952).
2. British Standard 1647 (1961).
3. Japanese Industrial Standard Z8802 (1958). Revised 1964.
4. R.G.Bates, *Determination of pH*, John Wiley and Sons, New York (1964).
5. R.G.Bates and B.A.Guggenheim, *Pure Appl. Chem.*, **1**, 163 (1960).
6. R.G.Bates, M.Paabo and R.A.Robinson, *J. Phys. Chem.*, **67**, 1833 (1963).
7. L.G.Van Uitert and C.G.Haas, *J. Am. Chem. Soc.*, **75**, 451 (1953).
8. A.G.Mitchell and W.F.K. Wynne-Jones, *Trans. Faraday Soc.*, **51**, 1690 (1955).
9. C.L.deLigny, P.F.M. Luykx, M.Rehbach and A.A.Wiencke, *Rec. Trav. Chim.*, **79**, 699,713(1960).
10. K.C.Ong, R.A.Robinson and R.G.Bates, *Analyt. Chem.*, **36**, 1971 (1964).
11. B.Gutbezahl and E.Grunwald, *J. Am. Chem. Soc.*, **75**, 565 (1953). E.F.Sieckmann and E.Grunwald, *J. Am. Chem. Soc.*, **76**, 3855 (1954).
12. I.T.Oiwa, *Sci. Rept., Tohoku Univ., First Ser.*, **41**, 129 (1957).
13. V.V.Alexsandrov and N.A.Izmailov, *Zhur. Fiz. Khim.*, **32**, 404 (1958). N.A.Izmailov, *Dokl. Akad. Nauk, S.S.S.R.*, **127**, 104 (1959).
14. M.Paabo, R.A.Robinson and R.G.Bates, *J. Am. Chem. Soc.*, **87**, 415 (1965).
15. R.Gary, R.G.Bates, and R.A.Robinson, *J. Phys. Chem.*, **63**, 3806 (1964); see also a paper by the same authors in *J.Phys. Chem.*, **69**, 2750 (1965).

11. RECOMMENDED SYMBOLS FOR SOLUTION EQUILIBRIA*

11.1 INTRODUCTION

The IUPAC-Chemical Society Tables entitled "Tables of Stability Constants of Metal-Ion Complexes" (*Chem. Soc. Special Publication*, No. 17, Second Edition, 1964) seem to have brought with them, even without any official decision by the IUPAC, a certain *de facto* standardization of symbols for equilibrium constants. Stability constants are increasingly used instead of instability constants, and the symbols K_n and β_n are being used for step-wise and cumulative constants in practically all papers published by the leading schools of coordination chemistry.

The Commission thinks that it is desirable to recommend standard symbols for equilibrium constants. However, it is hard to predict in which direction research will expand in decades to come, and hence the standards must not be too rigidly fixed to the perhaps rather limited interests of the majority of present-day equilibrium chemists.

Even authors following the recommendations below are strongly advised to define the symbols they are using in a prominent place in the beginning of each book or research paper.

11.2 GENERAL RULES

The general symbol for an *equilibrium constant* is K. The equilibrium constant for any reaction may be denoted by K, followed by the reaction formula within parentheses.

11.2.1 THERMODYNAMIC DEFINITION

Equilibrium constants are strictly thermodynamically defined in terms of activities; the (relative) activity for each species is defined in terms of some measurable quantity — e.g. a concentration, a partial pressure, a mole fraction — so that the activity approaches to this quantity as the system approaches a certain limiting state. For dissolved species the limiting state, the *medium*, is sometimes defined as the pure solvent (H_2O, C_2H_5OH, etc.). It may, however, equally be defined as a mixed solvent or a salt medium such as 3M $NaClO_4$ at $25°C$.

For a reaction involving only dissolved species, one may define the equilibrium constant as the limiting value of the concentration quotient when the concentrations of the reactants approach zero.

Examples

$K(HA \rightleftharpoons H^+ + A^-) = \lim ([H^+][A^-][HA]^{-1})$, as $[H^+] \to 0$, $[A^-] \to 0$, and $[HA] \to 0$ in the pure medium

* Based on the approved Recommendations published in *Pure and Applied Chemistry*, Vol. 18 (1969), pp. 459 - 464.

$$K(\text{HgBr}_2 + \text{Br}^- \rightleftharpoons \text{HgBr}_3^-) = \lim\,([\text{HgBr}_3^-][\text{HgBr}_2]^{-1}[\text{Br}^-]^{-1}), \text{ as } [\text{HgBr}_2] \to 0,$$
$[\text{Br}^-] \to 0$, and $[\text{HgBr}_3^-] \to 0$ in the pure medium.

The medium may then, for instance, be pure H_2O, 30% C_2H_5OH, or $3M\text{LiNO}_3$.

It is not often realised that equilibrium constants that refer to an *ionic medium* are as well thermodynamically defined as those referring to pure H_2O. The difference is only that another activity scale is being used, where e.g. $a(\text{Br}^-)/|\text{Br}^-| \to 1$ as $|\text{Br}^-| \to 0$ in the pure ionic medium, instead of in $H_2O(l)$.

The approximation activity = concentration has usually proved quite good, if the solute concentrations are only a small fraction of the medium ions, and is used by most of those who are determining equilibrium constants in ionic media, without any attempted extrapolation. The error involved in this approximation can as a rule be expected to be no greater than the uncertainty in extrapolation to zero concentration in pure H_2O.

11.2.2 NECESSARY SPECIFICATION

The formula must give the reacting species, and the phases they occur in. It may be stated in the text that, if nothing is said to the contrary, a species is in one specific phase; in the examples given, the temperature and pressure (if not 1 atmosphere) must be stated explicitly, and also the activity scale used. The pressure unit should thus be stated for gases. For dissolved substances it should be stated, for instance, whether the activities are defined so that the ratio (activity/concentration) approaches unity on dilution with pure solvent or with some ionic medium. The concentration unit should also be stated (M, mole/kg solvent, mole fraction).

11.2.3 SPECIAL AND *AD HOC* SYMBOLS

Special symbols are recommended below for a very few common types of equilibrium constants. These symbols must not be used for other purposes, if confusion is possible. Some of the recommended symbols may be a convenient shorthand writing in lists and tables but would nevertheless be too unwieldy for use in equations. Each author will surely invent convenient symbols such as K', K'', $K_{(8)}$, Q_3, etc. for his special purpose. Such *ad hoc* symbols must be clearly defined in some conspicuous place in the beginning of the text, and provided with the necessary specifications mentioned above.

11.3 COMPLEX FORMATION EQUILIBRIA

11.3.1 GENERAL SYMBOLS

The *overall formation constant* for any complex can be denoted by the symbol β followed, within parentheses, by the formula of the complex or, if there may be ambiguity, by the formulae of the components so that a formula for the complex is obtained by addition.

Examples

$$\beta(\text{FeSCNCl}_2) = K(\text{Fe}^{3+} + \text{SCN}^- + 2\text{Cl}^- \rightleftharpoons \text{FeSCNCl}_2)$$

This can also be written as $\beta(\text{Fe}^{3+}\text{SCN}^-(\text{Cl}^-)_2$, or $\beta(\text{Fe}^{3+}, \text{SCN}^-, 2\text{Cl}^-)$

$$\beta(\text{Th}(\text{HL})_2^{2+}) = \beta(\text{Th}^{4+}(\text{HL}^-)_2) = \beta(\text{Th}^{4+}, 2\text{HL}^-) = K(\text{Th}^{4+} + 2\text{HL}^- \rightleftharpoons \text{Th}(\text{HL})_2^{2+})$$

However,

$$\beta(\text{Th}^{4+}(H_2L)_2(H^+)_{-2}) = \beta(\text{Th}^{4+}, 2H_2L, -2H^+) = K(\text{Th}^{4+} + 2H_2L \rightleftharpoons \text{Th}(\text{HL})_2^{2+} + 2H^+)$$

11.3.2 MONONUCLEAR BINARY COMPLEXES

If a central atom (central group) M (the "metal") and a ligand L have been defined, then K_n is the *stepwise formation constant*, and β_n is the *cumulative formation constant* for the complex ML_n. They can both be referred to as stability constants (stepwise and cumulative):

$$K_n = K(ML_{n-1} + L \rightleftharpoons ML_n)$$
$$\beta_n = K(M + nL \rightleftharpoons ML_n)$$

Example For $M = Hg^{2+}$, $L = Cl^-$, $K_3 = \lim([HgCl_3^-][HgCl_2]^{-1}[Cl^-]^{-1})$;
$\beta_3 = \lim([HgCl_3^-][Hg^{2+}]^{-1}[Cl^-]^{-3})$

Equilibrium between a "metal" M and a protonated ligand HL, with liberation of H^+, is so often studied that it may sometimes be practical to use a special symbol. In "Stability Constants" the symbols $*K_n$ and $*\beta_n$ are used.

$$*K_n = K(M_{n-1} + HL \rightleftharpoons M + H^+)$$
$$*\beta_n = K(M + nHL \rightleftharpoons ML_n + nH^+)$$

(if $L = OH^-$, $HL = H_2O$).

Note: The use of M, L and n should be given preference in cases with simple mono-nuclear metal-ligand complexes, but other symbols for the reagents, such as A and B, and other symbols for the coefficients, such as $p, q, r, s...$, may be used if needed, e.g. in cases where one has several different ligands or several different metal atoms, or does not wish to distinguish between "metal" and "ligand" in the reactions.

11.3.3 POLYNUCLEAR AND MIXED COMPLEXES

For polynuclear complexes or complexes with several kinds of ligands it may sometimes be practical to use β with double or multiple subscripts. Their general meaning must then be defined very clearly with a full reaction formula. The examples given below of β's with a double subscript are only illustrations of convenient systems of notation but *the Commission does not find a strict standardization in this field necessary*.

For polynuclear complexes, the classification of one reagent as the "metal" and the other as the "ligand" may sometimes seem arbitrary. "Stability Constants", which had to choose one and be consistent, uses the definition:

$$\beta_{nm} = K(nL + mM \rightleftharpoons M_m L_n) ; \qquad K_{1n} = K(M_{n-1}L + M \rightleftharpoons M_n L)$$

By this definition, the second subscript gives the number of "metal" ions. For mononuclear complexes, $\beta_{n1} = \beta_n$ so that the second index can be dropped.

Examples

$$L = OH^-, \quad M = Sn^{2+} ; \quad \beta_{43} = K(3Sn^{2+} + 4OH^- \rightleftharpoons Sn_3(OH)_4^{2+}) ;$$
$$L = PO_4^{3-}, \quad M = H^+ ; \quad \beta_{13} = K(3H^+ + PO_4^{3-} \rightleftharpoons H_3PO_4) ;$$
$$K_{13} = K(H_2PO_4^- \rightleftharpoons H_3PO_4).$$

For mixed complexes one may use similar notations. The alphabetic order of the ligands is recommended if there is no strong reason for any other order.

Examples

$$\text{Ligands} = Br^-, I^- ; M = Sn^+ ; \beta_{rs} =$$

$$= K(Bi^{3+} + rBr^- + sI^- \rightleftharpoons BiBr_rI_s^{(3-r-s)+})$$

If there is no risk that, for instance, β_{22} could be read as "$\beta_{sub\ 22}$" then it can be written as here, otherwise a comma should be inserted, thus, $\beta_{2,2}$.

11.3.4 PROTONATION EQUILIBRIA

Equilibria with protons may be treated as a special case of complex formation; to make the treatment analogous with that of metal ion complexes, it is usually logical to make $H^+ = M$, the central atom. This is done consistently in the Tables of Stability Constants, and is illustrated above for $L = PO_4^{3-}$, $M = H^+$. Some special symbols are convenient, in addition:

11.3.4.1 The *acidity constant* K_{an} is the equilibrium constant for splitting off the nth proton from a charged or uncharged acid, to be defined. One may write K_a for K_{a1}.

11.3.4.2 The *protonation constant* K_{Hn} is the equilibrium constant for the addition of the nth proton to a charged or uncharged ligand, to be defined. The cumulative protonation constant β_{Hn} is the equilibrium constant for the formation of H_nL from nH^+ and L. For the ionization constant of water, the special symbol K_w may be used, $K_w = K(H_2O \rightleftharpoons H^+ + OH^-)$.

The same equilibrium constant may thus be described in several ways, dependent on how the ligand, L, or the acid, has been defined. This may be permissible because of the great importance of protonation equilibria.

Example

$$K(H^+ + H_2PO_4^- \rightleftharpoons H_3PO_4) \quad \text{may be denoted by}$$

K_{13} (if $L = PO_4^-$, $M = H^+$), by $K_{H3}(PO_4^{3-})$, or $K_H(H_2PO_4^-)$

or by the general symbol $\beta(H^+H_2PO_4^-)$. The inverted value

$$K(H_3PO_4 \rightleftharpoons H_2PO_4^- + H^+) \text{ is } K_a(H_3PO_4)$$

11.3.5 CONNECTED QUANTITIES (*AVERAGE LIGAND NUMBERS*, ETC.)

A barred letter — following the usage in other fields — denotes the average value for the quantity in question, among the species considered.

Examples

$$\bar{n} = \Sigma n[ML_n]/(\Sigma[ML_n]); \quad \bar{p} = \Sigma p[A_pB_q]/(\Sigma[A_pB_q]) ;$$

$$\bar{q} = \Sigma q[A_pB_q]/(\Sigma[A_pB_q])$$

The sums are, in general, taken for all the species ML_n with $n = 0,1,2...n_{max}$; and for all species A_pB_q except A and B. If some other species should be included or excluded in the sums, this must be clearly stated in the text.

The average number of L bound per M present can often be measured directly. If only mono-nuclear complexes are present, it is identical with $\bar{n}$, the average value of n in the formula ML_n.

If, however, polynuclear complexes are present and if their general formula is written as M_mL_n, then the average number of ligands bound per M is not the same as $\bar{n}$, the average value of n, and must be given another symbol, such as Z.

$$Z = \frac{\Sigma n |M_m L_n|}{\Sigma m |M_m L_n|} \qquad \bar{n} = \frac{\Sigma n |M_m L_n|}{\Sigma |M_m L_n|}$$

($|M|$ is included in the sums for Z, and may be included in those for $\bar{n}$). For instance, in a solution which contains practically only the complex M_2L_3, we would have $\bar{n} \sim 3$ (average value of n) but $Z \sim 1.5$ (average value of number of L bound per M).

What has been called Z might still be called $\bar{n}$ provided the complexes were written as $(ML_n)_m$ instead of $M_m L_n$, or if another letter than n were used in the general formula. Less confusion would arise, however, if another symbol than $\bar{n}$ were used in systems known to involve polymeric species.

11.4 SOLUBILITY EQUILIBRIA

For equilibria in which one solid phase is dissolved to give a number of species in solution one may use the shorthand notation, K_s, followed, within parentheses, by the formulae for the other participants in the reaction.

Examples

$$K_s(Ag^+, Ag(CN)_2^-) = K(2AgCN(s) \rightleftharpoons Ag^+ + Ag(CN)_2^-)$$

$$K_s(HgI_4^{2-}(I^-)_{-2}) = K_s HgI_4^{2-}, -2I^- = K(HgI_2(s) + 2I^- \quad HgI_4^{2-})$$

$$K_s(Cd^{2+}H_{-1}Cl^-) = K_s(Cd^{2+}, -H^+, Cl^-) = K(CdOHCl(s) + H^+ \rightleftharpoons Cd^{2+} + Cl^- + H_2O)$$

$$K_s(Fe^{3+}OH_{2.7}^- Cl_{0.3}^-) = K_s(Fe^{3+}, 2.7OH^-, 0.3Cl^-)$$

$$= K(Fe(OH)_{2.7}Cl_{0.3}(s) \rightleftharpoons Fe^{3+} + 2.7OH^- + 0.3Cl^-)$$

11.5 LIQUID-LIQUID DISTRIBUTION EQUILIBRIA

11.5.1 REACTION FORMULAE, DENOTATION OF PHASES

Typical distribution equilibria are the following:

$$HgCl_2 \text{ (in } H_2O) \rightleftharpoons HgCl_2 \text{ (in } C_6H_6)$$

$$UO_2^{2+} + 2NO_3^- \rightleftharpoons UO_2(NO_3)_2 \text{ (in } Et_2O)$$

$$Th^{4+} + 4HL(org) + 2T(org) \rightleftharpoons ThL_4T_2(org) + 4H^+$$

In the last case, one would have to define HL, T, and the solvent "org".

Most distribution equilibria studied at present involve (like those above) an aqueous phase and an organic phase. In the future, however, other combinations of phases may be studied extensively: SO (l), molten salts and metals, etc., and so it does not seem wise to restrict the symbolism to the organic/aqueous case.

The phases may be denoted by the formula of the solvent, or by "org" (to be defined in the text, or by overlining formulae referring to one phase, usually the less polar phase, or by Roman numerals (also to be defined in the text). One may state (as above) that species without phase notation are in one phase (here aqueous solution). To make typing and printing easier it is recommended that full formulae for solvents be given on the line, and not as a subscript, whereas Roman numerals may be used as subscripts, both in reaction formulae and in expressions for concentrations such as $[ThL_4T_2]_{II}$.

11.5.2 EQUILIBRIUM CONSTANTS

The general symbol for an equilibrium constant involving solutions in two phases is K_D. The following are examples of possible notations:

Examples

$$K_D(HgCl_2)_{II/I} = K(HgCl_2(I) \rightleftharpoons HgCl_2(II))$$

$$K_D(UO_2(NO_3)_2)_{org} = K(UO_2^{2+} + 2NO_3^- \rightleftharpoons UO_2(NO_3)_2(org)) \quad , \quad org = Et_2O$$

$$K_D(Th^{4+}H^+_{-4}\overline{HL}_4\overline{T}_2) = K(Th^{4+} + 4HL + 2T \rightleftharpoons \overline{ThL_4T_2} + H^+)$$

$$K_D(Th^{4+}H^+_{-4}(HL_{II})_4)_{II} = K(Th^{4+} + 4HL_{II} \rightleftharpoons ThL_4(II) + 4H^+)$$

This way of writing might imply a certain economy of space, especially where it is necessary to write out the formulae for L, T, and II.

The first type of equilibrium constant may be termed a *distribution constant* straight away and the others *overall distribution constants*.

11.5.3 CONNECTED QUANTITIES

The analytical *distribution ratio* for any component between two phases will be denoted by D, if necessary with the component and the phases stated, for instance $D(Hg)_{II/I}$, or D_{Hg}, or D. As a rule, D varies with the composition of the solution, as distinguished from K_D, which is a true equilibrium constant.

The symbol K with or without a subscript should only be used for true equilibrium constants, and the occasional misuse of any K symbols for variable concentration quotients must be discouraged.

12. RECOMMENDED NOMENCLATURE FOR LIQUID-LIQUID DISTRIBUTION

12.01   LIQUID-LIQUID DISTRIBUTION

This is the process of transferring a dissolved substance from one liquid phase to another (immiscible) liquid phase; the corresponding method of separation and concentration.

Comment.   A synonymous term for liquid-liquid distribution is *partition between two liquids*. The terms *solvent extraction* or *liquid extraction* are not recommended.

12.02   LIQUID-LIQUID EXTRACTION

This is a special case of liquid-liquid distribution and the term can be used where it is more appropriate.

12.03   EXTRACTION CONSTANT, K_{ex}

The extraction constant is the equilibrium constant for the distribution reaction. At zero ionic strength it is expressed as K_{ex}^{o}.

Comment.   For example, in the gross reaction

$$M_w^{n+} + nHL_{org} \rightleftharpoons ML_{n\ org} + nH_w^+ \qquad 12.1$$

(w = aqueous phase,   org = organic phase)

in which the reagent HL initially dissolved in the organic phase reacts with a metal ion in aqueous solution to form a product, ML_n, which is more soluble in the organic phase than in water, the equilibrium constant may be written as

$$K_{ex} = \frac{[ML_n]_{org}[H^+]_w^n}{[M^{n+}]_w[HL]_{org}^n} \qquad 12.2$$

*Based on the recommended nomenclature approved and published in *Pure and Applied Chemistry*, Vol.21, No.1 (1970), pp.109 - 114. This terminology has since been re-examined by Commission V.3 and greatly extended to include some 20 additional terms used in pure and applied sciences and in the technological applications of liquid-liquid extraction.

The distinction between the distribution constant (K_D) and the partition constant (K_D^o) and the (concentration) distribution ratio (D_c) is reaffirmed and it is recommended that the terms partition constant, partition coefficient, and extraction constant should not be used as synonyms for the (analytical) distribution ratio, D_c. A distinction is drawn between the terms *solvent* and *diluent*, and the term *extractant* is now restricted to the active substance in the solvent (i.e. the homogeneous 'organic' phase which comprises the *extractant*, the *diluent*, and/or the *modifier*) which is primarily responsible for the transfer of solute from the 'aqueous' to the 'organic' phase.

These recommendations, which are more in keeping with current practice in this rapidly developing field, will be published shortly in the form of a Provisional Nomenclature Appendix which will not as yet have received the full approval of IUPAC.

The phases can also be specified by the formula of the solvent, or by the subscript o or org (to be defined in the text) or by overlining formulae referring to one phase, usually the less polar phase. The subscript w (or aq) for the aqueous phase is usually omitted.

When the reagent is more soluble in water than in the other immiscible liquid it may be more convenient to define a special extraction constant in terms of $[HL]_w$. In distribution equilibria involving only non-aqueous systems (e.g. $SO_2(l)$, molten salts and metals) the mass action constant for the relevant process can be defined with K_{ex}.

12.04 DISTRIBUTION CONSTANT, K_D

The distribution constant is the ratio of the concentration of a substance in a single definite form in the organic solvent phase to its concentration in the same form in the aqueous phase at equilibrium; frequently termed a partition coefficient.

$$(K_D)_A = [A]_o/[A]_w \qquad 12.3$$

Comment. The use of the inverse concentration ratio (aqueous/organic) or the ratio of the concentration of the less dense phase to the concentration of the denser phase is not recommended.

12.05 PARTITION CONSTANT, K_D^o

This is the value of K_D attained as $[A] \to 0$.

12.06 (CONCENTRATION) DISTRIBUTION RATIO, D_C

This is the ratio of the total analytical concentration of a substance in the organic phase to its total analytical concentration in the aqueous phase, usually measured at equilibrium. (The descriptive adjective *concentration* can be omitted when there is no ambiguity with the Mass Distribution Ratio, D_m.)

Comment. The term *distribution* or *extraction coefficient* can be used in place of the term distribution ratio. However, the terms *partition constant* or *partition coefficient* should not be used in this connection since there has been confusion in the past between its use to describe a distribution ratio which varies with experimental conditions (e.g. pH, presence of complexing agents) and a true partition constant, K_D^o, which is by definition invariable (cf. 12.05), or a distribution constant, K_D, which is constant for a particular set of conditions (cf. 11.5.3).

12.07 RECOVERY FACTOR, R

The recovery factor is the fraction or percentage ($R_\%$) of the total quantity of a substance extracted (usually into the organic solvent phase) under specified conditions. Specifically $R_A = Q_A/(Q_A)'$, where $(Q_A)'$ and Q_A are the original and final quantities of the substance A.

Comment. The term extractability is not recommended. If an aqueous solution is extracted with n successive portions of organic solvent, the ratio of the volumes of the phases being $V_o/V_w = r$ in each case, the recovery factor for a particular substance is given by

$$R_n = 1 - (rD_C + 1)^{-n} \qquad 12.4$$

If $n = r = 1$, $R_1 = D_C/(1 + D_C)$.

12.08 ENRICHMENT FACTOR, S

The enrichment factor is that factor by which the original ratio of the concentrations of two substances to be separated must be multiplied to give the ratio after separation (usually in the organic phase).

$$\frac{Q_B}{Q_A} = S_{B/A} \frac{(Q_B)'}{(Q_A)'}, \qquad 12.05$$

Hence

$$S_{B/A} = Q_B(Q_A)'/Q_A(Q_B)' = R_B/R_A \qquad 12.6$$

Comment. In any analytically useful separation, $R_A \sim 1$, so that $S_{B/A} \sim R_B$. In terms of D, n and r, the enrichment factor is given by

$$S_{B/A} = \frac{1 - (1 + rD_B)^{-n}}{1 - (1 + rD_A)^{-n}} \qquad 12.7$$

If $n = 1$ and $r = 1$, $S_{B/A} = D_B(1 + D_A)/D_A(1 + D_B)$.

Comment. In liquid-liquid extraction distributions the enrichment factor is not given by D_B/D_A as often stated.

12.09 EXTRACTANT

A liquid phase (usually an organic solvent or the solution of an extracting agent in an organic solvent or solvent mixture) which is used to extract a substance from another liquid phase (usually aqueous).

Comment. This term should not be applied to one of the components of the extracting phase.

12.10 DILUENT

An inert (organic) solvent used to improve the physical properties (density, viscosity, etc.) or the extractive properties (e.g. selectivity) of the extractant.

12.11 EXTRACTING AGENT

The reagent which forms a complex salt or other adduct which partitions across the phase boundary of the extraction system.

12.12 EXTRACT

The separated phase (usually organic) containing the substance extracted from the other phase.

12.13 STRIPPING (BACK EXTRACTING)

The process of back-extraction of the substance from the extract (usually into an aqueous phase).

12.14 STRIPPING SOLUTION

A solution (usually aqueous, sometimes water alone) used for extracting the substance from the extract (usually organic).

12.15 SCRUBBING

The process of removing impurities from the separated phase containing the main substance (i.e. from the extract or *back-extract*).

12.16 SCRUBBING SOLUTION

An aqueous or organic solution used for scrubbing.

12.17 SALTING-OUT

Improving the extraction of a substance by the addition of an electrolyte to the aqueous phase.

Comment. The term salting-out is used here in a wider sense than it is in some other fields.

13. RECOMMENDATIONS ON NOMENCLATURE AND PRESENTATION OF DATA IN GAS CHROMATOGRAPHY*

13.1 INTRODUCTION

Gas chromatography is so widely used that it has become necessary to standardize the definitions and presentation of results. Recommendations having this objective should be in harmony with gas-chromatographic theory and account has been taken of this in the following proposals.

Gas chromatography is almost always carried out by elution. The recommendations pertaining to the presentation of results and general background (Section 13.5 et seq.) are restricted to elution, mainly gas-liquid, chromatography; further recommendations will be required for gas-solid chromatography. The definitions of terms (Sections 13.2-13.4) are nearly all generally applicable without restriction.

13.2 NAME OF TECHNIQUE

13.2.01 GAS CHROMATOGRAPHY

Gas chromatography comprises all chromatographic methods in which the moving phase is a gas. The word chromatography itself implies that a stationary phase is present in addition to the moving phase.

13.2.02 GAS-LIQUID CHROMATOGRAPHY

Gas-liquid chromatography comprises all gas-chromatographic methods in which the stationary phase is a liquid distributed on a solid support. Separation is achieved by partition of the components of a sample between the phases.

13.2.03 GAS-SOLID CHROMATOGRAPHY

Gas-solid chromatography comprises all gas chromatographic methods in which the stationary phase is an active solid (e.g. charcoal, molecular sieves). Separation is achieved by adsorption of the components of a sample.

13.3 APPARATUS

13.3.01 SAMPLE INJECTOR

A sample injector is a device by which a liquid or gaseous sample is introduced into the apparatus. The sample can be introduced directly into the carrier-gas stream, or into a

---

* These Recommendations are based on the approved report published in *Pure and Applied Chemistry*, Vol. 8 (1964), pp. 553 - 562, which revised and updated the Preliminary Recommendations published in *Pure and Applied Chemistry*, Vol. 1 (1960), pp. 177 - 185.

chamber temporarily isolated from the system by valves which can be changed so as to make an instantaneous switch of the gas stream through the chamber. The latter is a *by-pass injector*.

13.3.02 SOLID VOLUME

Solid volume is the volume occupied by the solid support or the active solid in the column.

13.3.03 LIQUID VOLUME, V_L

The liquid volume is that volume occupied by the liquid phase in the column. $V_L = w_L/\rho_L$, where w_L is the weight of the liquid in the column, and ρ_L is its density at the column temperature.

13.3.04 INTERSTITIAL VOLUME, V_G

The interstitial volume is the volume of the column not occupied by the liquid phase and its solid support, or by the active solid. It does not include any volume external to the column, such as the volume of the sample injector or of the detector.

13.3.05 DETECTOR

A detector is a device that measures the change of composition of the effluent. A detector that measures instantaneous concentration is called a *differential detector*. An *integral detector* continuously measures the sample accumulated from the beginning of the analysis.

13.4 REAGENTS

13.4.01 CARRIER GAS

Carrier gas (or eluent gas) is gas used to elute the sample as it passes through the column. The carrier gas together with the portions of the sample present in this phase constitutes the *mobile phase*.

13.4.02 LIQUID PHASE

The liquid phase is a liquid which is relatively non-volatile at the column temperature and is sorbed on the solid support, where it acts as a solvent for the sample. Separation depends on differences in solubility of the various components of the sample in the liquid phase.

13.4.03 SOLID SUPPORT

The solid support is normally an inert porous solid, which sorbs the liquid phase. The particle-size range of the support affects column efficiency and the pressure differential necessary to achieve a given flow rate. Modifications of the method have been introduced for the achievement of special separations, in which the solid support is not inert but is an active solid. In capillary columns the inner wall of the column serves as the solid support and obviates the use of additional porous solids for this purpose.

13.4.04 ACTIVE SOLID

The active solid is a porous solid with adsorptive properties by means of which chromatographic separations may be achieved. The separations resulting from this action follow laws different from those deriving from the partitioning action of the liquid phase.

13.4.05 STATIONARY PHASE

In gas-liquid chromatography the *stationary phase* comprises the liquid phase without the solid support. In gas-liquid chromatography the *stationary phase* is the active solid.

13.5 CHROMATOGRAPHIC RECORDS

13.5.01 CHROMATOGRAM

A chromatogram is a plot of the detector response against time or the volume of the carrier gas. Idealized chromatograms obtained with differential and integral detectors for one component are shown in Figure 13.6.

The definitions in this paragraph apply to the chromatograms obtained directly by means of differential detectors or by differentiating the records obtained by means of integral detectors. The *base line* is that portion of a chromatogram recorded when only carrier gas emerges from the column.

13.5.02 PEAK

A peak is the portion of a chromatogram recording the detector response while a single component emerges from the column. (If separation of a mixed sample is incomplete, two or more components may appear as one peak). The *peak base* CD is an interpolation of the base line between the extremities of the peak. The area enclosed between the peak and the base line is the *peak area* and the distance BE from the peak maximum to the peak base measured parallel to the axis representing detector response is the *peak height*. The segment of peak base FG intercepted by tangents to the inflection points on either side of the peak is the *peak width*. The line parallel to the peak base, bisecting the peak height, and terminating at the sides of the peak HJ, is the *peak width at half height*.

13.5.03 STEP (AND STEP HEIGHT)

The following definitions apply to chromatograms obtained with integral detectors. As a sample component passes through the detector, a sigmoid curve is obtained and the base line is displaced to a new position. The change in base-line position caused by the sample component is known as a *step*, and the difference in the heights of the two base lines is the *step height*.

13.6 RETENTION PARAMETERS

13.6.01 RETENTION VOLUME (UNCORRECTED)

The retention volume (uncorrected), V_R, is the volume of gas required to elute the compound under study and is given by

$$V_R = t_R F_c \qquad 13.6.1$$

where t_R is the *retention time*, the time for the emergence of the peak maximum after injection of the sample, and F_c is the volumetric flow rate of the carrier gas measured at the outlet pressure and temperature of the column. V_R, t_R correspond to OB in Figure 13.6 which, in the remaining definitions, is assumed to have the carrier-gas volume plotted as the horizontal axis.

13.6.02 GAS HOLD-UP, V_M

The gas hold-up, V_M, is the uncorrected retention volume of a non-absorbed sample and is the volume of carrier gas required to transport such a sample from the point of injection to the point of detection at column outlet pressure. It includes contributions due to the interstitial volume of the column and the effective volumes of the sample injector and the detector. It can readily be determined for any column by elution of some material for which the partition coefficient is very small compared with its value for other solutes.

Gases such as nitrogen, air or the noble gases are normally employed for this purpose. The peak often produced by the presence of small amounts of air during the sample injection gives this information, and is referred to as the *air peak*.

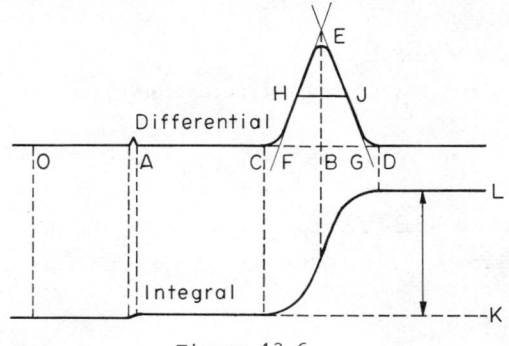

Figure 13.6

For a capillary column the interstitial volume may be calculated from the dimensions. The interstitial volume divided by j (see below) is the contribution to V_M due to the column and the contribution due to the apparatus may therefore be determined.

13.6.03 ADJUSTED RETENTION VOLUME, V_R'

The adjusted retention volume, V_R', is given by

$$V_R' = V_R - V_M = AB \qquad \qquad 13.6.2$$

13.6.04 CORRECTED RETENTION VOLUME, V_R^o

The corrected retention volume, V_R^o, is given by

$$V_R^o = jV_R = j(OB) \qquad \qquad 13.6.3$$

This quantity is of limited use because it is influenced by the volumes of the sample injector and detector as well as the interstitial volume of the column.

13.6.05 PRESSURE-GRADIENT CORRECTION FACTOR, j

The symbol j in equation 13.6.3 is the *pressure-gradient correction factor* for a homogeneously filled column of constant diameter and is given by

$$j = \frac{3}{2}\left[\frac{(p_i/p_0)^2 - 1}{(p_i/p_0)^3 - 1}\right] \qquad \qquad 13.6.4$$

where p_i, p_o are the pressures of the carrier gas at the inlet and the outlet of the column respectively. Use of the factor j allows for the fact that in gas chromatography the mobile phase is compressible. If in fact the flow rate is measured at the inlet of the column, the corrected retention volume may be obtained by using a suitably modified expression for j.

The correction factor j should strictly be applied only to parameters which relate to the column volume alone and are unaffected by the volumes of the injector and detector. The retention volume (= $V_R' + V_G/j$) referring to an ideal chromatographic apparatus in which the volumes of the injector and detector are zero may be called the theoretical retention volume. For most purposes there is no need to evaluate the theoretical retention volume but the definition is included here in case the distinction is needed for didactic or theoretical purposes.

13.6.06 NET RETENTION VOLUME, V_N

The net retention volume, V_N, is given by

$$V_N = jV_R' = j(AB) \qquad 13.6.5$$

13.6.07 SPECIFIC RETENTION VOLUME, V_g

The specific retention volume, V_g, is the net retention volume at $0°C$ per gram of liquid phase and is given by

$$V_N/w_L = V_g T/273 \qquad 13.6.6$$

where T is the absolute temperature of the column. V_N/w_L is the net retention volume per gram at the column temperature. A suitable symbol for this quantity is V_g^θ.

13.6.08 RELATIVE RETENTION, $r_{A/B}$

The relative retention, $r_{A/B}$, is the retention volume for the compound under study expressed relative to the retention volume of a standard (reference) compound on the same column at the same temperature. If the subscripts A and B refer to the substance and reference compound respectively,

$$r_{A/B} = \frac{V_{g,A}}{V_{g,B}} = \frac{V_{N,A}}{V_{N,B}} = \frac{V_{R,A}'}{V_{R,B}'} \neq \frac{V_{R,A}}{V_{R,B}} \qquad 13.6.7$$

Relative retentions measured from the point of injection can only be considered independent of column dimensions if $V_M \ll V_{R,A}$ and $V_{R,B}$. When, as is usual and desirable, relative retentions are determined from one and the same chromatogram in which experimental conditions are constant and identical for both components, the determination is simplified to the measurement of the appropriate distances on the recorder chart (i.e. the distance corresponding to the adjusted retention volumes).

13.6.09 PARTITION COEFFICIENT, K

The partition coefficient, K, is defined as

$$K = \frac{(\text{weight of solute})/(\text{cm}^3 \text{ stationary phase})}{(\text{weight of solute})/(\text{cm}^3 \text{ mobile phase})}$$

and is assumed to be independent of concentration at the concentrations prevailing in gas chromatography.

According to elementary theory, which has been adequately verified by experiment, the partition coefficient is related to the retention volume by

$$K = \frac{V_N}{V_L} = \frac{V_N \rho_L}{w_L} = \frac{V_g T \rho_L}{273} \qquad 13.6.8$$

The specific retention volume, the relative retention, and the partition coefficient are independent of column parameters, but they do depend upon the samples involved, the partitioning system, and the temperature.

13.6.10 MEANING OF QUALIFYING SIGNS

In definitions of retention parameters the superscript $°$ indicates that the pressure correction factor has been applied, and the prime $'$ that measurements are made from the air peak. However, the symbol for net retention volume with this scheme is unduly cumbersome, and V_N has been substituted for it.

13.7 RECOMMENDATIONS: RETENTION DATA

Measurements of retention data should be reported in such a manner that they can be converted for use in experiments with other apparatus and under different conditions. This can be done on an absolute basis, by measurement of the partition coefficient or specific retention volume; or on a relative basis, by measurement of relative retentions, relative to a standard solute. For determining the relative retentions of a series of substances a standard should be chosen such that its retention volume falls near the middle of the series. Standards with very small retention volumes should not be used.

13.7.01 TEMPERATURE EFFECTS

Whenever possible the variation of retention volume with temperature should be found and results reported for at least two temperatures, as far apart as possible. If results are sufficiently extensive, a suitable graphical method is to plot the logarithm of the relative retention against the reciprocal of the absolute temperature. Variation of specific retention volume may be expressed in a similar way by means of an Antoine equation:

$$\log V_g = A + \frac{B}{t + C} \qquad 13.7.1$$

in which t is the column temperature in $°C$ and A, B, C, are constants. The relationships so obtained can conveniently be used for interpolation.

13.7.02 EXPERIMENTAL DETAILS

The following experimental variables should be published with any set of results laying claim to being quantitative in nature:

> nature and particle-size range of solid support;
> nature, concentration, and amount of liquid phase in solution;
> sample size;
> column dimensions (length and internal diameter);
> column inlet and outlet pressures;
> flow-rate of carrier gas and method of measurement;
> temperature of column and accuracy of temperature control;
> description of detector, e.g. type of sensing element, cell geometry,
> cell volume, response time.

13.8 APPARATUS PERFORMANCE

13.8.01 COLUMN PERFORMANCE

An expression for column performance in terms of the theoretical plate number n can be calculated from the expression

$$n = 16 \times \left[\frac{\text{retention volume}}{\text{peak width}}\right]^2 = 16\left[\frac{OB}{FG}\right]^2 \qquad 13.8.1$$

(see Figure 13.6 above). The theoretical plate number may vary with the compound as well as the column. Therefore the compound used should be specified. The units for retention and peak width used in equation 13.8.1 must be consistent, so that their ratio is of dimension unity. If the corrected retention volume is used, the observed peak width must also be corrected for the pressure drop in the column.

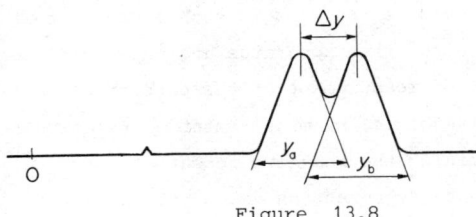

Figure 13.8

13.8.02 PEAK RESOLUTION

If two compounds are well enough separated to permit a satisfactory estimation of the peak width, and the peaks are approximately Gaussian, as shown in Figure 13.8, the resolution may be expressed by

$$\text{Resolution} = 2 \times \frac{\text{difference between retention volumes}}{\text{sum of peak widths}} \qquad 13.8.2$$

$$= 2\Delta y / (y_a + y_b)$$

13.9 DISCUSSION

13.9.01 GENERAL CONSIDERATIONS

The partition coefficient for a given solute-solvent system is (for the conditions prevailing in gas-liquid chromatography) a physical constant dependent only on the temperature, and gas-liquid chromatography provides a convenient method for its determination. The specific retention volume V_g has the same character of a general constant, and can easily be converted to K by means of the relationship 13.6.8. In the determination of K it is necessary to determine the density of the solvent at the column temperature (to about 1 per cent) while this is not necessary for the determination of V_g. The other column variables and operating conditions, however, have to be accurately known since they enter into the computation of K and V_g, as can be seen from the relationships given above.

In the determination of relative retentions, it is not necessary to know any column variables (e.g. F_c, w, p_i, p_o), except the temperature; all that is necessary is that they remain constant. Furthermore, relative retentions do not vary with temperature as much as do absolute measurements, and are therefore to be preferred unless the variables listed can be determined with accuracy. Relative retentions, used with standard substances, as suggested above, are immediately useful for the identification of compounds if tables of retentions including the compounds in question are available.

It is important to specify the ratio of liquid phase to solid support precisely. The activity of the latter can be such as to influence appreciably the chromatographic separations achieved; this effect will be more pronounced the lower the amount of liquid covering the solid.

13.9.02 ERRORS

The following factors can affect the retention parameters and will cause errors unless they are corrected for: sample size, method of injection, and detector dead volume. These factors affect not only the retention parameters but also the peak shape, and therefore can give misleading results also in the calculation of efficiency and resolution. These calculations, then, should not be relied upon unless the distorting factors are small, and the peaks obtained nearly Gaussian.

13.9.03 EXPERIMENTAL CONSIDERATIONS

The flow rate, F_c, is required at the temperature and outlet pressure of the column, whereas measurements of flow are usually made at room temperature. Suitable corrections must therefore be made: if a capillary flowmeter is used, the pressure drop across the meter must be considered; with wet flowmeters allowance must be made for the vapour pressure of the water. If F is the flow rate of the saturated gas determined from the flowmeter at pressure p, p_w is the vapour pressure of water at the temperature of the flowmeter, T_c is the temperature (K) of the column and T_m is the temperature of the flowmeter, the partial pressure of the carrier gas, p_M, is given by

$$p_M = p - p_w \qquad 13.9.1$$

and the flow rate by

$$F_c = F(p - p_w) \, T_c / p T_m \qquad 13.9.2$$

The carrier gas should enter the column at column temperature; the sample should be made to vaporize very rapidly on injection in order to avoid artefacts of efficiency or resolution.

In absolute measurements, account must be taken of the temperature of operation in assessing the life of a column before a significant change in w_L occurs. The rate of variation of the partition coefficient with temperature is similar in magnitude to that of vapour pressure, and the accuracy of temperature control, both with time and along the column, needs to be specified.

13.9.04 NOTES ON THE ALTERATIONS FROM THE PRELIMINARY RECOMMENDATIONS OF 1960

The term theoretical retention volume (Section 13.6.06) has now been introduced because the retention volume, V_R, as defined from the point of injection, includes the effective volumes of the sample injector and the detector.

The symbol V_g^θ has been introduced because of the strong preference expressed by many gas chromatographers for the use of this quantity in place of V_g.

Some amendments have been made in the French and German equivalents in the following Table of Terms.

13.10 TABLE OF TERMS

English	French	German	
Gas chromatography	Chromatographie des gaz Chromatographie en phase gazeuse	Gas-Chromatographie	
Gas-liquid chromatography	Chromatographie gaz-liquide	Flüssigkeits-Gas-Chromatographie	
Gas-solid chromatography	Chromatographie gaz-solide	Festkörper-Gas Chromatographie	
Sample injector	Injecteur d'échantillon	Einlass-System	
By-pass injector	Injecteur à dérivation	Umleit-Probengeber	
Differential detector	Détecteur différentiel	Differentialdetektor	
Integral detector	Détecteur intégral	Integraldetektor	
Solid volume	Volume solide	Festkörpervolumen	
Liquid volume	Volume liquide	Flüssigkeitsvolumen	
Interstitial volume	Volume interstitiel	Gasvolumen der Säule	V_o
Carrier gas	Gaz porteur, gaz vecteur	Trägergas	
Mobile phase	Phase mobile	mobile Phase	
Stationary phase	Phase stationnaire	stationäre Phase	
Liquid phase	Phase liquide	flüssige Phase, Trennflüssigkeit	
Solid support	Support solide	Träger	
Active solid	Solide actif	Absorbens	
Chromatogram	Chromatogramme	Chromatogramm	
Base line	Ligne de base	Null-Linie	
Peak	Pic	peak	
Peak base	Base du pic	peak-Basis	
Peak area	Surface du pic	peak-Fläche	
Peak height	Hauteur du pic	peak-Höhe	
Peak width	Largeur du pic	peak-Breite	
Peak width at half height	Largeur du pic à demi-hauteur	Halbwertsbreite	
Step	Palier	Stufe	
Step height	Hauteur de palier	Stufenhöhe	
Retention volume	Volume de rétention	Durchbruchsvolumen	V_R
Adjusted retention volume	Volume de rétention réduit	reduziertes Retentionsvolumen	V'_R
Corrected retention volume	Volume de rétention limite	korrigiertes Retentionsvolumen	V^o_R
Net retention volume	Volume de rétention absolu	Netto- Retentionsvolumen	V_N
Specific retention volume	Volume de rétention spécifique	spezifisches Retentionsvolumen	V_g
Pressure-gradient correction factor	Facteur de correction du gradient de pression	Faktor für die Druckkorrektion	j
Gas hold-up	Retenue de gaz	Totvolumen	V_M
Relative retention	Rétention relative	relative Retention	
Partition coefficient	Coefficient de partage	Verteilungs-Koeffizient	K
Column performance	Efficacité de la colonne	Trennschärfe der Säule	

TABLE 13.10 (*Continued*)

English	French	German	
Peak resolution	Résolution des pics	Auflösung	
Number of theoretical plates	Nombre de plateaux théoriques	Zahl der theoretischen Stufen	n

REFERENCES

D.H.Desty (Ed.). *Vapour Phase Chromatography 1956*, p.xi, Butterworths, London (1957).

D.H.Desty (Ed.). *Gas Chromatography, 1958*, p. xi, Butterworths, London (1958).

D. Ambrose, A.I. Keulemans, and J.H.Purnell, *Analyt. Chem.*, 30, 1582 (1958).

H.W.Johnson and F.H.Stross, *Analyt. Chem.*, 30, 1586 (1958).

V.J.Coates, H.J.Noebels, and I.S.Fagerson (Ed.). *Gas Chromatography*, p. 315, Academic Press, New York (1958).

J.Buzon, P.Chovin, L.Fanica, R.Ferrand, G.Guiochon, M.Huguet, J.Lebbe, J.Serpinet, and J.Tranchaut, *Bull. Soc. Chim. France*, 1137 (1959).

E.Kovats, *Helv. Chim. Acta*, 41, 1951 (1958).

Preliminary Recommendations on Nomenclature and Presentation of Data in Gas Chromatography, *Pure Appl. Chem.*, 1, 177 (1960).

M.B.Evans and J.F.Smith, *J. Chromatog.*, 6, 293 (1961).

Glossary of Terms Relating to Gas Chromatography, British Standard 3282. British Standards Institution, London (1963).

14. RECOMMENDATIONS ON NOMENCLATURE FOR CHROMATOGRAPHY*

14.1 CHROMATOGRAPHY

A method used primarily for separation of the components of a sample in which the components are distributed between two phases, one of which is stationary while the other moves. The stationary phase may be a solid, or a liquid supported on a solid, or a gel. The stationary phase may be packed in a *column*, spread as a *layer*, or distributed as a *film*, etc.; in these definitions *chromatographic bed* is used as a general term to denote any of the different forms in which the stationary phase may be used. The mobile phase may be gaseous or liquid.

14.2 PRINCIPAL METHODS

14.2.1 FRONTAL CHROMATOGRAPHY
A procedure for chromatographic separation in which the sample (liquid or gas) is fed continuously into the chromatographic bed.

14.2.2 ELUTION CHROMATOGRAPHY
A procedure for chromatographic separation in which an *eluent* (see Section 14.8.6) is passed through the chromatographic bed after the application of the sample.

14.2.3 DISPLACEMENT CHROMATOGRAPHY
An elution procedure in which the *eluent* contains a compound more effectively retained than the components under examination.

14.3 CLASSIFICATION ACCORDING TO PHASES USED

In this classification the first word specifies the mobile phase and the second the stationary phase. A liquid stationary phase is supported on a solid.

14.3.1 GAS CHROMATOGRAPHY (GC)

14.3.1.1 *Gas-liquid chromatography* (GLC)

14.3.1.2 *Gas-solid chromatography* (GSC)

---

* These recommendations are reproduced from the approved report published in *Pure and Applied Chemistry*, Vol. 37, No. 4 (1974), pp. 445-462.

In gas chromatography the distinction between gas-liquid and gas-solid may be obscure because liquids are used to modify solid stationary phases, and because the solid supports for liquid stationary phases affect the chromatographic process. For classification by the phases used, the term relating to the predominant effect should be chosen.

14.3.2   LIQUID CHROMATOGRAPHY (LC)

14.3.2.1 *Liquid-liquid chromatography* (LLC)

14.3.2.2 *Liquid-solid chromatography* (LSC)

14.3.2.3 *Liquid-gel chromatography* (LGC)

Liquid-gel chromatography includes gel-permeation and ion-exchange chromatography.

14.4 CLASSIFICATION ACCORDING TO MECHANISMS

14.4.1   ADSORPTION CHROMATOGRAPHY

Separation based mainly on differences between the adsorption affinities of the components for the surface of an active solid.

14.4.2   PARTITION CHROMATOGRAPHY

Separation based mainly on differences between the solubilities of the components in the stationary phase (gas chromatography), or on differences between the solubilities of the components in the mobile and stationary phases (liquid chromatography).

14.4.3   ION-EXCHANGE CHROMATOGRAPHY

Separation based mainly on differences in the ion-exchange affinities of the components.

14.4.4   PERMEATION CHROMATOGRAPHY

Separation based mainly upon exclusion effects, such as differences in molecular size and/or shape (e.g. molecular-sieve chromatography) or in charge (e.g. ion-exclusion chromatography). The term *gel-permeation chromatography* is widely used for the process when the stationary phase is a swollen gel. The term *gel-filtration* is not recommended.

14.4.5   OTHER MECHANISMS

In addition to the above mechanisms (Sections 14.4.1 to 14.4.4) there exist many techniques based upon other mechanisms. Examples are ligand-exchange, formation of charge-transfer complexes, and bio-specific sorption, e.g. formation of enzyme-substrate and antigen-antibody complexes. Classification according to mechanism should be avoided unless the predominant mechanism is known. In many instances more than one mechanism is involved.

14.5 CLASSIFICATION ACCORDING TO TECHNIQUES USED

All types of chromatography can be classified according to Section 14.3 by the phases used or according to Section 14.4 by mechanism, but the terms in this section specify techniques and may proved a more useful characterization of the process.

14.5.1   COLUMN CHROMATOGRAPHY (CC)

14.5.2   OPEN-TUBE CHROMATOGRAPHY (see Section 14.8.4)

14.5.3   PAPER CHROMATOGRAPHY (PC)

14.5.4   THIN-LAYER CHROMATOGRAPHY

Chromatography carried out in a layer of adsorbent spread on a support, e.g. a glass plate.

14.5.5   FILAMENT CHROMATOGRAPHY

14.6   TERMS FOR SPECIAL TECHNIQUES

14.6.1   TEMPERATURE-PROGRAMMED CHROMATOGRAPHY

A procedure in which the temperature of the column is changed systematically during a part or the whole of the separation.

14.6.2   FLOW-PROGRAMMED CHROMATOGRAPHY

A procedure in which the rate of flow of the mobile phase is changed systematically during a part or the whole of the separation.

14.6.3   SALTING-OUT CHROMATOGRAPHY

A procedure in which a non-sorbable electrolyte is added to the eluent to modify the distribution equilibria of the components to be separated.

14.6.4   SELECTIVE ELUTION

An elution procedure in which a specific eluent is used, e.g. a complexing agent that forms stable non-sorbable complexes with one or a group of the compounds to be separated, but affects the other components only to a negligible extent.

14.6.5   STEPWISE ELUTION

An elution procedure in which two or more eluents of different composition are used in succession to elute the components in a single chromatographic run.

14.6.6   GRADIENT ELUTION

An elution procedure in which the composition of the eluent is changed continuously during a single chromatographic run.

14.6.7   TWO-DIMENSIONAL CHROMATOGRAPHY

A procedure applied in paper chromatography and in thin-layer chromatography in which the components are caused to migrate first in one direction, and subsequently in a direction at right angles to the first one. The two elutions are generally carried out with different eluents.

14.6.8   REVERSED-PHASE CHROMATOGRAPHY

A term of historical interest in liquid-liquid chromatography referring to an elution procedure in which the stationary phase is non-polar, e.g. paper treated with hydrocarbons or silicones.

14.7   TERMS RELATING TO THE METHOD IN GENERAL

14.7.01   CHROMATOGRAM

A graphical or other representation of detector response, effluent concentration, or other quantity used as a measure of effluent concentration versus time or the volume of the effluent. Idealized chromatograms for one component, obtained with differential or integral detectors, are shown in Figure 14.7 (a) and (b). The term chromatogram is also applied to a chromatographic paper, layer, or column after separation has occurred.

14.7.02 ELUTION CURVE
A chromatogram or part of a chromatogram, recorded when elution techniques are used.

14.7.03 CHROMATOGRAPH (verb)
To separate by chromatography.

14.7.04 CHROMATOGRAPH (noun)
The assembly of apparatus for carrying out chromatographic separations.

14.7.05 ELUTE
To chromatograph by elution chromatography. This term is preferred to the term *develop*, which has been used in paper chromatography and in thin-layer chromatography. The process of elution may continue until the components have left the chromatographic bed.

14.7.06 EXTRACT
To recover a compound from a chromatographic zone by treatment with a solvent.

14.7.07 ZONE
A region in a chromatographic column or layer where one or more components of the sample are located.

14.7.08 SPOT
A zone in paper or thin-layer chromatography of approximately circular appearance.

14.7.09 STARTING POINT OR LINE
The point or line on a chromatographic layer where the substance to be chromatographed is applied.

14.7.10 BASELINE
The portion of a chromatogram recorded when only eluent or carrier gas emerges from the column.

14.7.11 PEAK
The portion of a differential chromatogram (see Section 14.8.20) recording the detector response or eluate concentration (see Section 14.8.18) while a single component emerges from the column (Figure 14.7). If separation is incomplete, two or more components may appear as one *unresolved peak*.

14.7.12 ELUTION BAND
Synonymous with peak.

14.7.13 TAILING
Asymmetry of a peak such that, relative to the base line, the front is steeper than the rear. In paper chromatography and in thin-layer chromatography, the distortion of a zone showing a diffuse region behind the zone in the direction of travel.

14.7.14 FRONTING
Asymmetry of a peak such that, relative to the baseline, the front is less steep than the rear. In paper chromatography and thin-layer chromatography, the distortion of a zone showing a diffuse region in front of the zone in the direction of flow.

14.7.15 STEP (on an integral chromatogram)

The portion of an integral chromatogram (see Section 14.8.21) recording the amount of a component, or the corresponding change in the signal from the detector as the component emerges from the column (cf. Figure 14.7).

14.7.16 STEP HEIGHT (on an integral chromatogram)

The distance (KL in Figure 14.7), perpendicular to the time or volume axis, through which the baseline moves as a result of a step on an integral chromatogram (see Section 14.8.21).

14.7.17 INTERNAL STANDARD

A compound added to the sample in known concentration, for example, for the purpose of eliminating the need to measure the size of sample in quantitative analysis.

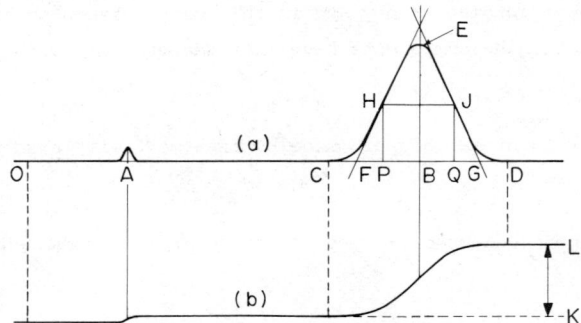

Figure 14.7 (a) Differential, and (b) integral chromatograms

14.7.18 MARKER

A reference substance chromatographed with the sample to assist in identifying the components.

14.8 TERMS RELATING TO THE SEPARATION PROCESS AND THE APPARATUS

14.8.01 COLUMN

The tube that contains the stationary phase, and through which the mobile phase passes.

14.8.02 PACKING

The active solid, stationary liquid plus solid support, or swollen gel put in the column. The term *packing* refers to the conditions existing before the chromatographic run is started (i.e. to the material introduced into the column) whereas the *stationary phase* (see Section 14.8.08) refers to the conditions during the run.

14.8.03 PACKED COLUMN

A column filled with packing.

14.8.04 OPEN TUBULAR COLUMN

A column, usually of capillary dimensions, in which the column wall, a liquid or an active solid supported on the column wall acts as the stationary phase.

14.8.05 MOBILE PHASE

The phase that is moving in the chromatographic bed. It includes the fraction of the sample present in this phase.

14.8.06 ELUENT
The liquid or gas entering the chromatographic bed and used to effect a separation by elution.

14.8.07 CARRIER GAS
The term normally used for the *eluent* in gas chromatography.

14.8.08 STATIONARY PHASE
The non-mobile phase in the chromatographic bed, on which the separation depends. For example, in gas-solid chromatography and liquid-solid chromatography the active solid is the stationary phase, and in gas-liquid and liquid-liquid chromatography the liquid, but not the solid support, is the stationary phase.

14.8.09 ACTIVE SOLID
A solid with sorptive properties by means of which chromatographic separations may be achieved.

14.8.10 MODIFIED ACTIVE SOLID
An active solid, the adsorptive properties of which have been changed by treatment with a gas, liquid, or another solid.

14.8.11 SOLID SUPPORT
A solid that holds the stationary liquid phase.

14.8.11.1 SOLID VOLUME
The volume occupied by the solid support.

14.8.12 SUPPORT PLATE
The plate that supports the thin layer in thin-layer chromatography.

14.8.13 GRADIENT LAYER OR GRADIENT PACKING
A layer or column packing with continuous change of property affecting the separation, e.g. a pH gradient.

14.8.14 SAMPLE INJECTOR
A device by means of which a sample is introduced into the eluent (carrier gas) or the column.

14.8.15 BYPASS INJECTOR
A sample injector by means of which the eluent (carrier gas) may be temporarily diverted through a sample chamber so that the sample is carried to the column.

14.8.16 CHAMBER SATURATION
Uniform distribution of the eluent vapour throughout the chamber prior to chromatography.

14.8.17 LAYER EQUILIBRATION
Saturation of the stationary phase with the mobile phase via the vapour phase.

14.8.18 ELUATE
The effluent from a chromatographic bed emerging when elution is carried out.

14.8.19 DETECTION
The process by which the presence of chromatographically separated substances is recognized.

14.8.20 DIFFERENTIAL DETECTOR

A detector whose response is dependent on the instantaneous difference in composition between the column effluent and the eluent (carrier gas).

14.8.21 INTEGRAL DETECTOR

A detector whose response is dependent on the total amount of a component that has passed through it.

14.8.22 SOLVENT FRONT

The front line of the eluent.

14.8.23 SOLVENT MIGRATION-DISTANCE

The distance travelled by the solvent front.

14.8.24 SEPARATION TEMPERATURE

The temperature of the chromatographic bed: often the column temperature in column chromatography.

14.8.25 INJECTION TEMPERATURE

The temperature at the point of injection.

14.8.26 INITIAL AND FINAL TEMPERATURES

The range of separation temperatures in temperature-programmed chromatography.

14.9 TERMS RELATING TO QUANTITATIVE EVALUATION AND THE THEORY OF CHROMATOGRAPHY

14.9.01 COLUMN VOLUME, X

The volume (empty) of the part of the column that contains the packing. It is recommended that the column dimensions be given as the inner diameter and the height or length of the column occupied by the stationary phase under the applied chromatographic conditions. If swelling changes occur, the conditions under which the height is determined should be specified.

14.9.02 BED VOLUME

Synonymous with *column volume* for a packed column.

14.9.03 INTERSTITIAL VOLUME, V_I

The volume occupied by the mobile phase in the packed section of a column. In gas chromatography the gas occupying the interstitial volume expands to a volume V_I/j at the outlet pressure, where measurements are normally made (see Section 14.9.11).

14.9.04 INTERSTITIAL FRACTION, ε_I

The interstitial volume per unit volume of a packed column:

$$\varepsilon_I = V_I/X \qquad 14.9.1$$

14.9.05 VOLUME OF THE STATIONARY PHASE, V_S

The volume of the stationary liquid phase or of the active solid or of the gel in the column. The volume of any solid support is not included.

14.9.06 STATIONARY-PHASE FRACTION, ε_S

The volume of the stationary phase per unit volume of a packed column

$$\varepsilon_S = V_S/X \qquad 14.9.2$$

14.9.07 PHASE RATIO

The ratio of the volume of the mobile phase to that of the stationary phase in a column.

14.9.08 HOLD-UP VOLUME, V_M

The volume of eluent required to elute a component the concentration of which in the stationary phase is negligible compared to that in the mobile phase. The *hold-up volume* corresponds to the distance OA in Figure 14.7, and includes any volumes contributed by the sample injector and the detector.

14.9.09 GAS HOLD-UP VOLUME, V_M

Synonymous with the *hold-up volume* in gas chromatography. The volume of carrier gas (eluent) is specified at the same temperature and pressure as the total retention volume (see Section 14.9.23).

14.9.10 DEAD VOLUME, V_d

The volume between the effective injection point and the effective detection point, less the column volume X.

14.9.11 PRESSURE-GRADIENT CORRECTION-FACTOR, j

A factor, applying to a homogeneously filled column of uniform diameter, that corrects for the compressibility of the mobile phase; the values of the measured quantities obtained after multiplication by the factor j are independent of the pressure drop in the column. In practice, these quantities include contributions arising from the column inhomogeneities making up the dead volume but since these are small in comparison with retention volumes the consequent errors are normally ignored. In gas chromatography, if p_i, p_o are respectively the pressures of the carrier gas at the inlet and outlet of the column:

$$j = \frac{3\left[(p_i/p_o)^2 - 1\right]}{2\left[(p_i/p_o)^3 - 1\right]} \qquad 14.9.3$$

14.9.12 PEAK BASE

The interpolation, in a differential chromatogram, of the baseline between the extremities of the peak (the line CD in Figure 14.7).

14.9.13 PEAK AREA

The area (CHEJD in Figure 14.7) enclosed between the peak and the peak base.

14.9.14 PEAK MAXIMUM

The point on the peak at which the distance to the peak base, measured in a direction parallel to the axis representing the detector response, is a maximum (E in Figure 14.7).

14.9.15 PEAK HEIGHT

The distance between the peak maximum and the peak base, measured in a direction parallel to the axis representing the detector response (the distance BE in Figure 14.7).

14.9.16 PEAK WIDTH

The segment of the peak base intercepted by tangents to the inflection points on either side of the peak (the distance FG in Figure 14.7) projected onto the axis representing time or volume if the baseline is not parallel to this axis.

14.9.17 PEAK WIDTH AT HALF HEIGHT

The length of the line parallel to the peak base that bisects the peak height and terminates at the intersections with the two limbs of the peak (the distance PQ in Figure 14.7) projected onto the axis representing time or volume if the baseline is not parallel to this axis.

14.9.18 VOLUMETRIC FLOWRATE, F_c

The volumetric flowrate of the mobile phase (cm^3 min^{-1}). In gas chromatography, the flowrate is normally specified at the column temperature and outlet pressure, although the measurement may be made at ambient temperature and must be corrected accordingly (and possibly also for water vapour present in the flowmeter).

14.9.19 NOMINAL LINEAR FLOW, F

The volumetric flowrate of the mobile phase divided by the area of the cross section of the column (cm min^{-1}), i.e. the linear flowrate in a part of the column not containing packing.

14.9.20 INTERSTITIAL VELOCITY, u (u_o at the outlet pressure in gas chromatography)

The linear velocity of the mobile phase inside a packed column calculated as the average over the entire cross section. This quantity can, under idealized conditions, be calculated from the equation :

$$u = F/\varepsilon_I \qquad 14.9.4$$

14.9.21 MEAN INTERSTITIAL VELOCITY OF THE CARRIER GAS, $\bar{u}$

The interstitial velocity of the carrier gas multiplied by the pressure-gradient correction-factor:

$$\bar{u} = Fj/\varepsilon_I \qquad 14.9.5$$

14.9.22 RETENTION VOLUMES

Retention measurements (and measurements of hold-up volume and peak width) may be made in terms of times, e.g. t_R, t'_R analogous to V_g, V'_g (see Sections 14.9.23 and 14.9.25), or chart distances as well as volumes. If flow and recorder speeds are constant, the volumes are directly proportional to the times and chart distances. The definitions given here are drawn up in terms of volume, and it is recommended that theoretical discussion should be couched in the same terms wherever possible. However, the proportionality between volumes, times, and chart distances is implied in references to Figures 14.7 and 14.9.

14.9.23 TOTAL RETENTION VOLUME, V_R

The volume of eluent (carrier gas) entering the column between the injection of the sample and the emergence of the peak maximum of the specified component (OB in Figure 14.7). It includes the hold-up volume. In gas chromatography the volume of carrier gas is specified at the outlet pressure and temperature of the column.

Note: The word *total* in this definition allows *retention volume* to be used as a general term when specification of a particular quantity is not required.

14.9.24 PEAK ELUTION VOLUME, $\overline{V}$

The volume of eluent entering the column between the start of the elution and the emergence of the peak maximum. The term applies only to liquid chromatography. It does not include the effluent obtained when the sample is introduced into the column nor the volume of the detector, if used.

Sometimes the column is washed with a liquid, before the elution is started, but after application of the sample, to displace components that are not retained. The effluent obtained during this washing process is not included in the peak elution volume unless the solutes are moved during the washing (see Section 14.6.5).

14.9.25 ADJUSTED RETENTION VOLUME, V_R'

The total retention volume less the hold-up volume (corresponding to the distance AB in Figure 14.7), i.e.,

$$V_R' = V_R - V_M = \overline{V} - V_I \qquad 14.9.6$$

14.9.26 NET RETENTION VOLUME, V_N

The adjusted retention volume multiplied by the pressure-gradient correction-factor:

$$V_N = jV_R' \qquad 14.9.7$$

14.9.27 SPECIFIC RETENTION VOLUME, V_g

The net retention volume per gram of stationary liquid, active solid or solvent-free gel. In liquid chromatography, except when conducted at very high pressures, the compression of the mobile phase is negligible, and the adjusted and net retention volumes are identical; the specific retention volume is then the adjusted retention volume per gram of stationary liquid, active solid, or solvent-free gel. It is recommended that, when appropriate, authors specify the drying conditions. At $0°C$, $V_g = 273\, V_N/w_L T$, where w_L is the mass of the stationary liquid phase.

14.9.28 RELATIVE RETENTION, $r_{A/B}$

The adjusted retention volume of a substance relative to that of a reference compound obtained under identical conditions. If subscripts A and B refer to the substance and the reference compound respectively, then

$$r_{A,B} = \frac{V_{g,A}}{V_{g,B}} = \frac{V_{N,A}}{V_{N,B}} = \frac{V_{R,A}'}{V_{R,B}'} \qquad 14.9.8$$

Comment: Note that $r_{A/B}$ is not equal to $V_{R,A}/V_{R,B}$ nor to $\overline{V}_A/\overline{V}_B$.

14.9.29 RETENTION TEMPERATURE

The column temperature (see Section 14.8.24) when the peak maximum for a component has been reached in temperature-programmed chromatography.

14.9.30 R_f VALUE

The ratio of the distance travelled by the centre of a zone to the distance simultaneously travelled by the mobile phase. In paper and thin-layer chromatography, R_f may be determined from the distance moved by the eluent front.

14.9.31 R_B VALUE

The ratio of the distance travelled by a zone to the distance simultaneously travelled by a reference substance B.

14.9.32 DISTRIBUTION CONSTANT, K_D

The ratio of the concentration of a component in a single definite form in the stationary phase to its concentration in the same form in the mobile phase at equilibrium. Both concentrations are calculated per unit volume of the phase.

This term is recommended in preference to *partition coefficient*, which has been used with the same meaning.

In chromatography a component may be present in more than one form; these forms are generally not specified (and may not be known), and it will therefore usually be more appropriate for specification of conditions in the column to use one of the following terms, which are defined by the analytical concentration (or amount) of the component, the analytical concentration referring to its total concentration (or amount) without regard to its possible existence in associated or dissociated forms.

14.9.33 CONCENTRATION DISTRIBUTION RATIO, D_c

The ratio of the analytical concentration of a component in the stationary phase to its analytical concentration in the mobile phase:

$$D_c = \frac{\text{amount of component} / \text{cm}^3 \text{ of stationary phase}}{\text{amount of component}/ \text{cm}^3 \text{ of mobile phase}} \qquad 14.9.9$$

14.9.34 DISTRIBUTION COEFFICIENTS, D_g, D_v, D_s

The amount of a component in a specified amount of stationary phase, or in an amount of stationary phase specified by its surface area, divided by the analytical concentration in the mobile phase. The subscripts, g, v, and s, indicate as follows the way in which the stationary phase is specified.

$$D_g = \frac{\text{amount of component}/ \text{gram of dry stationary phase}}{\text{amount of component}/ \text{cm}^3 \text{ of mobile phase}} \qquad 14.9.10$$

applicable in ion-exchange and gel chromatography, where swelling occurs, and in adsorption chromatography with adsorbents of unknown surface area,

$$D_v = \frac{\text{amount of component in the stationary phase} /\text{cm}^3 \text{ of bed volume}}{\text{amount of component}/\text{cm}^3 \text{ of mobile phase}} \qquad 14.9.11$$

applicable when it is not practicable to determine the weight of the solid phase, and

$$D_s = \frac{\text{amount of component}/\text{m}^2 \text{ of surface}}{\text{amount of component}/\text{cm}^3 \text{ of mobile phase}} \qquad 14.9.12$$

14.9.35 MASS DISTRIBUTION RATIO, D_m

The fraction $(1 - R)$ of a component in the stationary phase divided by the fraction (R) in the mobile phase:

$$D_m = \frac{\text{amount of component in the stationary phase}}{\text{amount of component in the mobile phase}} \qquad 14.9.13$$

This term is recommended in preference to the term *capacity factor* frequently used in gas chromatographic literature.

The subscripts in D_c, D_m, D_g, D_s may be omitted when there is no possibility of confusion of one term with another.

Values of these quantities, defined in Sections 14.9.33 - 14.9.35, which allow the equilibrium between two phases to be specified, may be determined by static equilibrium measurements. They may also be related to retention volumes, and measurements of the latter frequently provide the most convenient experimental route for their determination.

14.9.36 SEPARATION FACTOR, $\alpha_{A/B}$

The ratio of the distribution ratios or coefficients D_A/D_B for two substances A and B measured under identical conditions. By convention α is usually greater than unity.

14.9.37 PEAK RESOLUTION, R_s

The separation of two peaks in terms of their average peak width (see Figure 14.9).

$$R_s = 2y/(y_A + y_B) \qquad 14.9.14$$

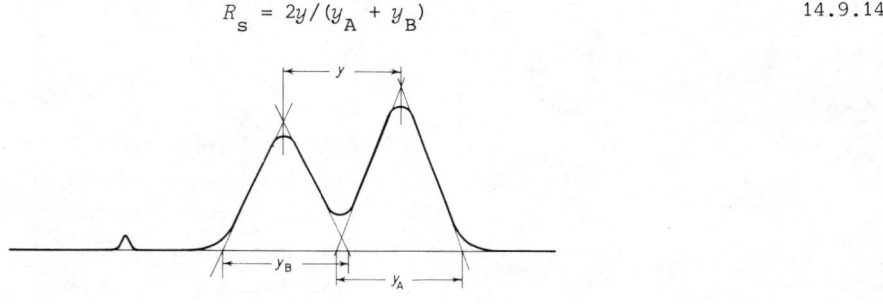

Figure 14.9

14.9.38 THEORETICAL PLATE NUMBER, n

A number indicative of column performance calculated from the equation

$$n = 16 \text{ (Peak elution volume/peak width)}^2 \qquad 14.9.15$$

In gas chromatography and some types of liquid chromatography the volumes of the sample injector and of the detector are negligible, and the expression for n can then be written in the form

$$n = 16 \text{ (Total retention volume/peak width)}^2 \qquad 14.9.16$$

In these expressions the units for the quantities inside the brackets must be consistent so that their ratio is of unit dimensions, i.e. if the numerator is a volume, then peak width must be expressed in terms of volume also.

14.9.39 EFFECTIVE THEORETICAL PLATE NUMBER, N

A number indicative of column performance when resolution is taken into account:

$$N = 16 \, R_s^2/(1 - \alpha)^2 \qquad 14.9.17$$

14.9.40 HEIGHT EQUIVALENT TO A THEORETICAL PLATE, HETP, h

The column length divided by the theoretical plate number.

14.9.41 HEIGHT EQUIVALENT TO AN EFFECTIVE THEORETICAL PLATE, HEETP, H

The column length divided by the effective theoretical plate number.

14.9.42 RETENTION INDEX, I

A number, obtained by logarithmic interpolation, relating the adjusted retention volume of a component A to the adjusted retention volumes of the normal paraffins. Each n-paraffin is arbitrarily allotted by definition an index one hundred times its carbon number. The index I_A of substance A is then given by

$$I_A = 100N + 100n \frac{\log V'_R(A) - \log V'_R(N)}{\log V'_R(N+n) - \log V'_R(N)} \qquad 14.9.18$$

where $V'_R(N+n)$ and $V'_R(N)$ are the adjusted retention volumes of n-paraffins of carbon numbers $(N+n)$ and N that are respectively smaller and larger than $V'_R(A)$, the adjusted retention volume of A.

14.10 APPENDIX. LIST OF SYMBOLS

Symbol	Section	Meaning
A, B	14.9.28	Components A and B
D_c	14.9.33	Concentration distribution ratio
D_g	14.9.34	Distribution coefficient
D_m	14.9.35	Mass distribution ratio
D_s	14.9.34	Distribution coefficient
D_v	14.9.34	Distribution coefficient
F	14.9.19	Nominal linear flow
F_c	14.9.18	Volumetric flowrate
H	14.9.41	Height equivalent to an effective theoretical plate, HEETP
h	14.9.40	Height equivalent to a theoretical plate, HETP
I	14.9.42	Retention index
j	14.9.11	Pressure-gradient correction-factor
K_D	14.9.32	Distribution constant
N	14.9.39	Effective theoretical plate number
n	14.9.38	Theoretical plate number
p_i, p_o	14.9.11	Pressure of carrier gas at inlet and outlet of the column
R	14.9.35	Fraction of component in the mobile phase
R_f	14.9.30	R_f value
R_B	14.9.31	R_b value
R_s	14.9.37	Peak resolution
$r_{A/B}$	14.9.28	Relative retention (of component A relative to component B)
u	14.9.20	Interstitial velocity
$\bar{u}$	14.9.21	Mean interstitial velocity
u_o	14.9.20	Interstitial velocity at the column outlet
$\bar{V}$	14.9.24	Peak elution volume
V_d	14.9.10	Dead volume

14.10 LIST OF SYMBOLS (continued)

Symbol	Section	Meaning
V_g	19.9.27	Specific retention volume
V_I	19.9.03	Interstitial volume
V_M	19.9.08, 19.9.09	Hold-up volume, gas hold-up volume
V_N	19.9.26	Net retention volume
V_R	19.9.23	Total retention volume
V'_R	19.9.25	Adjusted retention volume
V_S	19.9.05	Volume of the stationary phase
X	19.9.01	Column volume
$\alpha_{A/B}$	19.9.36	Separation factor
ε_I	19.9.04	Interstitial fraction
ε_S	19.9.06	Stationary-phase fraction

REFERENCES

1. Recommendations on Nomenclature and Presentation of Data in Gas Chromatography, *Pure Appl. Chem.*, 8, 553 (1964).
2. Recommended Nomenclature for Liquid-Liquid Distribution, *Pure Appl. Chem.*, 21, 111 (1970).
3. Recommendations on Ion Exchange Nomenclature, *Pure Appl. Chem.*, 29, 619 (1972).
4. Recommended Practice for Gas Chromatography Terms and Relationships, *ASTM* E-355-68, American Society for Testing and Materials: Philadelphia (1968); *J. Gas Chromatogr.*, 6, 1 (1968).
5. E. Bayer *et al.*, *Chromatographia*, 1, 153 (1969).
6. E. Stahl, *Chromatographia*, 1, 338 (1968).
7. *British Standards 3282, Glossary of Terms Relating to Gas Chromatography*, British Standards Institution: London (1969).
8. *Le Système International d'Unités*, Offilib; rue Gay-Lussac, F 75 Paris 5 (1970); *The Internation System of Units (SI)*, NBS Spec. Publ. 330, National Bureau of Standards; Washington DC (1972).
9. Manual of Symbols and Terminology for Physicochemical Quantities and Units, *Pure Appl. Chem.*, 21, 1 (1970).

15. RECOMMENDATIONS ON ION EXCHANGE NOMENCLATURE*

15.1 INTRODUCTION

The Commission has at all times been aware of the need to harmonize its recommendations with the existing recommended nomenclature for gas chromatography (Section 13), liquid-liquid distribution (Section 12) and other separation processes.

15.2 DEFINITIONS

15.2.01 ION EXCHANGER : A solid or liquid, inorganic or organic, containing ions exchangeable with others of the same sign present in a solution in which the exchanger is considered to be insoluble. †

15.2.02 ION EXCHANGE : The process of exchanging ions between a solution and an ion exchanger.

15.2.03 COUNTER-IONS : In an ion exchanger, the mobile exchangeable ions.

15.2.04 FIXED IONS : In an ion exchanger, the non-exchangeable ions which have a charge opposite to that of the counter-ions.

15.2.05 IONOGENIC GROUPS : In an ion exchanger, the fixed groupings which are either ionized or capable of dissociation into fixed ions and mobile counter-ions.

15.2.06 CO-IONS : In an ion exchanger, mobile ionic species with a charge of the same sign as the fixed ions.

15.2.07 CATION EXCHANGER : An ion exchanger with cations as counter-ions. The term cation-exchange resin may be used in the case of solid organic polymers.

15.2.08 ANION EXCHANGER : An ion exchanger with anions as counter-ions. The term anion-exchange resin may be used in the case of solid organic polymers.

---

* Based upon the approved Recommendations published in *Pure and Applied Chemistry*, Vol.29, No.4 (1972), pp. 617 - 624.

† It is recognized that there are cases where liquid exchangers are employed where it may be difficult to distinguish between the separation process as belonging to ion-exchange or liquid-liquid distribution, but the broad definition given here is regarded as that which is most appropriate.

15.2.09	RESIN MATRIX :	The molecular network of an ion exchanger which carries the ionogenic groups.
15.2.10	CATION EXCHANGE:	The process of exchanging cations between a solution and a cation exchanger.
15.2.11	ANION EXCHANGE:	The process of exchanging anions between a solution and an anion exchanger.
15.2.12	ACID FORM OF CATION EXCHANGER :	The ionic form of a cation exchanger in which the counter-ions are hydrogen ions (H-form) or the ionogenic groups have added a proton forming an undissociated acid.
15.2.13	BASE FORM OF ANION EXCHANGER :	The ionic form of an anion exchanger in which the counter-ions are hydroxide groups (OH-form) or the ionogenic groups form an uncharged base, e.g. $-NH_2$.
15.2.14	SALT FORM OF AN ION EXCHANGER :	The ionic form of an ion exchanger in which the counter-ions are neither hydrogen nor hydroxide ions. When only one valence is possible for the counter-ion, or its exact form or charge is not known, the symbol or the name of the counter-ion without charge is used, e.g. sodium form, Na-form, tetramethylammonium form, orthophosphate form. When one of two or more possible forms is exclusively present, the oxidation state may be indicated by Roman numerals, e.g. Fe(II)-form, Fe(III)-form.
15.2.15	MONOFUNCTIONAL ION EXCHANGER :	An ion exchanger containing only one type of ionogenic group.
15.2.16	BIFUNCTIONAL ION EXCHANGER :	An ion exchanger containing two types of ionogenic groups.
15.2.17	POLYFUNCTIONAL ION EXCHANGER :	An ion exchanger containing more than one type of ionogenic group.
15.2.18	MACROPOROUS ION EXCHANGER :	An ion exchanger with pores that are large compared to atomic dimensions.
15.2.19	COLUMN VOLUME, X :	The total volume of that part of a column which contains the ion exchanger. It is recommended that the column dimensions be given as the inner diameter and the height or length of the column occupied by the ion exchanger under the applied conditions. If swelling changes occur the conditions under which the height is determined should be specified. The dimensions should be given in mm or cm.
15.2.20	BED VOLUME :	Synonymous with column volume for a packed column.

15.2.21	THEORETICAL SPECIFIC CAPACITY, Q_o :	Milliequivalents of ionogenic group per gram of dry ion exchanger. If not otherwise stated the capacity should be reported per gram of the H-form of a cation exchanger and Cl-form of an anion exchanger.
15.2.22	VOLUME CAPACITY, Q_v :	Milliequivalents of ionogenic group per cm^3 (true volume) of swollen ion exchanger. (The ionic form of the ion exchanger and the medium should be stated.)
15.2.23	BED VOLUME CAPACITY :	Milliequivalents of ionogenic group per cm^3 of bed volume determined under specified conditions (should always be given together with specification of these conditions).
15.2.24	PRACTICAL SPECIFIC CAPACITY, Q_A :	Total amount of ions expressed in milliequivalents or millimoles taken up per gram of dry ion exchanger under specified conditions (should always be given together with the specification of these conditions).
15.2.25	BREAK-THROUGH CAPACITY OF ION EXCHANGER BED, Q_B:	The practical capacity of an ion exchanger bed obtained experimentally by passing a solution containing a particular ionic or molecular species through a column containing the ion exchanger, under specified conditions, and measuring the amount of species which has been taken up when the species is first detected in the effluent or when the concentration in the effluent reaches some arbitrarily defined value. The break-through capacity may be expressed in milliequivalents, millimoles or milligrams taken up per gram of dry ion exchanger or per cm^3 of bed volume.
15.2.26	WEIGHT SWELLING IN SOLVENT, w_s (e.g. w_{H_2O}):	Grams of solvent taken up by one gram of the dry ion exchanger.
15.2.27	VOLUME SWELLING RATIO :	Ratio of the dry swollen volume to the true dry volume.
15.2.28	SELECTIVITY COEFFICIENT, $k_{A/B}$:	The equilibrium coefficient obtained by formal application of the law of mass action to ion exchange and characterizing quantitatively the relative ability of an ion exchanger to select one of two ions present in the same solution. Exchange $Mg^{2+} - Ca^{2+}$ $k_{Mg/Ca} = [\overline{Mg}][Ca]/[Mg][\overline{Ca}]$ Exchange $SO_4^{2-} - Cl^-$ $k_{SO_4^{2-}/Cl^-} = [SO_4]_r[Cl]^2/[SO_4][Cl]_r^2$ Over-bars or subscript letters, 'r', are used to designate concentrations in the ion exchanger. For exchanges involving counter-ions differing in their charges, the numerical value of $k_{A/B}$ depends on the

choice of the concentration scales in the ion exchanger and the solution (molal scale, molar scale, mole fraction scale, etc.). Concentration units must be clearly stated in exchange of ions of differing charges.

15.2.29 CORRECTED SELECTIVITY COEFFICIENT, $k_{A/B}^a$: Concentrations of external solutions in 15.2.28 are replaced by activities.

15.2.30.1 CONCENTRATION DISTRIBUTION RATIO, *D_c: The ratio of the (total) analytical concentration of a solute in the ion exchanger to its analytical concentration in the external solution. The concentrations are calculated per cm^3 of the swollen ion exchanger and cm^3 of the external solution.

15.2.30.2 DISTRIBUTION COEFFICIENT, D_g* : The ratio of the total (analytical) amount of solute per gram of dry ion exchanger to its concentration (total amount per cm^3) in the external solution.

15.2.30.3 VOLUME DISTRIBUTION Coefficient*, D_v : The ratio of the total (analytical) amount of a solute in the ion exchanger calculated per cm^3 of column or bed volume to its concentration (total amount per cm^3) in the external solution. ($D_v = D_g \rho$, where ρ is the bed density expressed in grams of dry resin per cm^3 of bed.) This quantity is most conveniently determined from column experiments and it is recommended to use D_v values in describing the results from chromatographic separations.

15.2.31 SEPARATION FACTOR, $\alpha_{A/B}$: $\alpha_{A/B} = D_A/D_B$. The ratio between the distribution coefficients of solutes A and B in a specified medium at a specified temperature. In exchange of counter-ions of equal charge the separation factor is equal to the selectivity coefficient provided that only one type of ion represents the analytical concentration (e.g. in exchanges of K^+ and Na^+, but not in systems where several individual species are included in the analytical concentrations).

15.2.32 ION EXCHANGE ISOTHERM : The concentration of a counter-ion in the ion exchanger expressed as a function of its concentration in the external solution under specified conditions and at constant temperature.

15.2.33 SORPTION : Uptake of electrolytes or non-electrolytes by ion exchangers through mechanisms other than pure ion exchange.

15.2.34 SORPTION ISOTHERM : The concentration of a sorbed species in the ion

* Definitions, 15.2.30.1, 15.2.30.2 and 15.2.30.3 are used both for ions and for non-electrolytes.

exchanger expressed as a function of its concentration in the external solution under specified conditions and at constant temperature.

15.2.35 DIFFUSION COEFFICIENT, $\bar{D}$: The diffusion coefficient in the ion exchanger.

15.2.36 ION-EXCHANGE MEMBRANE : A thin sheet or film of ion-exchange material which may be used to separate two solutions and which allows the preferential transport of either cations (in the case of a cation-exchange membrane) or anions (in the case of an anion-exchange membrane). The membrane material may be made only from ion exchanging material, when it is called a *homogeneous ion-exchange membrane*, or the ion exchange material may be embedded in an inert binder and it is then called a *heterogeneous ion-exchange membrane*.

15.2.37 PERMSELECTIVITY : Permeation of certain ionic species in preference to other species through ion-exchange membranes.

15.2.38 REDOX POLYMERS : Polymers containing functional groups which can be reversibly reduced or oxidized. 'Electron exchanger' may be used as a synonym.

15.2.39 REDOX ION EXCHANGERS : Conventional ion exchangers in which reversible redox couples have been introduced as counter-ions or by sorption or complex formation. They closely resemble redox polymers in their behaviour.

The following terms have already been defined in the Recommendations on Nomenclature and Presentation of Data in Gas Chromatography (Section 13) and in the Recommendations on Nomenclature for Chromatography (Section 14) and are listed here for convenience.

15.2.40 RELATIVE RETENTIONS : $r_{A/B}$

15.2.41 ADJUSTED RETENTION VOLUME : V_R'

15.2.42 PEAK : (Elution band may be used synonymously)

15.2.43 PEAK BASE :

15.2.44 PEAK AREA :

15.2.45 PEAK WIDTH :

15.2.46 COLUMN PERFORMANCE :

15.2.47 PEAK RESOLUTION :

15.2.48 MOBILE PHASE :

16. NOMENCLATURE, SYMBOLS, UNITS AND THEIR USAGE IN SPECTROCHEMICAL ANALYSIS—I. GENERAL ATOMIC EMISSION SPECTROSCOPY*

16.1 FOREWORD

Many of these terms have already been defined in several nomenclature documents, especially those developed by IUPAC (*International Union of Pure and Applied Chemistry*), IUPAP (*International Union of Pure and Applied Physics*) and ASTM (*American Society for Testing Materials*). The fact that many of the symbols, units, nomenclature and definitions previously recommended are repeated in this document demonstrates that the nomenclature of a specific field, i.e. spectrochemical analysis, is deeply rooted in the general nomenclature of chemistry and physics. However, the adaptation of a general system to a specialized field requires a careful selection of general terms and the addition of new ones. In a few cases it was found necessary to deviate from symbols previously recommended in order to avoid using the same symbol for different quantities. Even in a restricted field, the same symbol may have different meanings, e.g. the letter c may stand for the speed of light or for concentration depending upon the context.

Some remarks about the arrangement of material in this document are appropriate. Some sections start with a list of terms and symbols, accompanied - if necessary - by short explanatory notes. This was done for general terms and symbols to facilitate reference. Other sections are in the form of a glossary of terms and definitions. This document is not therefore arranged in the systematic order of a textbook; it should be used as a compendium offering information at different levels and for different purposes.

16.2 GENERAL RECOMMENDATIONS AND PRACTICES

16.2.1 For the description of general quantities used in physics and chemistry, the nomenclature and symbols adopted in the most recent official documents of international scientific unions and organizations should be followed. The most important documents are:

1. 'Symbols, Units and Nomenclature in Physics'. *Document UIP 11 (SUN 65-3)*; International Union of Pure and Applied Physics [for short IUPAP 1965]: German edition: Friedr. Vieweg und Sohn, Braunschweig.

2. 'Manual of Physicochemical Symbols and Terminology' (IUPAC), *Pure Appl. Chem.*, 21, 1 (1970).

3. Publications of the International Organization for Standardization, Technical Committee 12 (ISO/TC 12).

16.2.2 The symbol for a physical quantity stands for the product of the numerical value

---

* Based on the approved Recommendations published in *Pure and Applied Chemistry*, Vol.30, No. 3 - 4 (1972), pp. 651 - 679.

(the measure), which is a pure number, and the unit:

$$\text{physical quantity} = \text{numerical value} \times \text{unit}$$

Therefore in equations composed of symbols for physical quantities, units should not appear.

16.2.3 Symbols for physical quantities should be single letters of the Latin or Greek alphabets printed in inclined or upright type, with or without modifying signs, i.e. subscripts, superscripts and dashes.
Symbols for units of physical quantities should be printed in upright type.
Numerals should be printed in upright type.
Symbols for chemical elements should be printed in upright type.
Indices that are symbols for physical quantities should be printed in inclined type.

(Rules from IUPAP 1965)

16.2.4 Following the recommendations of IUPAP 1965, a comma should be used in writing decimals, while in the English language texts a full stop [period] is permitted. Correspondingly, a cross (x) should be used for the multiplication sign but only between figures or numbers - not between symbols or units - while in texts not in English a point in the middle of the line is permitted.

Quantities less than unity expressed in decimal form should be written with a zero preceding the decimal sign. To help recognition of large numbers, the figures may be grouped in threes, with a [2-unit] space separating pairs of groups. Commas or full stops should not be used for this purpose.

16.2.5 In this document logarithms will all be understood to be logarithms to the base 10 (symbol: $\log x = \log_{10} x$).

16.2.6 Certain symbols and letters extensively employed in mathematics should be reserved for this purpose. These include: d, ∂ (partial differential); δ,Δ (difference); Σ (sum); f (function); < >, ⁻ (average), and x,y,z for spatial coordinates, and for the general description of measurable quantities.

16.2.7 Usage of certain general words in connection with numerical values.
The term *constant* should only be used for numerical values which really are constant, particularly for universal constants, such as the gas constant or the Boltzmann constant. 'Constant' may also be applied to unvarying material constants, such as dielectric constant. The term *coefficient* should only be used for numerical values which occur in equations of proportionality, for example, 'coefficient of recombination'. (In English these numbers are often indicated by the termination 'ity', e.g. 'absorptivity'.)

'*Index*' should only be used to indicate values arising from ratios, e.g. 'refractive index'. Since quantities having the dimension unity† are concerned, it is sufficient on occasion also to use the simple expression 'number'.

16.2.8 The 'International System of Units (SI)' is recommended. A summary of its base ‡ and some supplementary and derived units is given in Table 16.2.

    † Formerly called 'dimensionless'.

    ‡ The word 'base' was chosen by the Commission Générale des Poids et Mesures with respect to non-English languages.

TABLE 16.2

Summary of SI base, supplementary, and derived units

Physical quantity	Name of SI unit	Symbol for SI unit	Definition of units	Type
length	metre	m	–	Base
mass	kilogramme	kg	–	Base
time	second	s	–	Base
electric current	ampere	A	–	Base
thermodynamic temperature	kelvin	K	–	Base
luminous intensity	candela	cd	–	Base
amount of a substance	mole	mol	–	Base
plane angle	radian	rad	–	Supplm.
solid angle	steradian	sr	–	Supplm.
force	newton	N	$kg\ m\ s^{-2}$	Derived
pressure	pascal	Pa	$kg\ m^{-1}\ s^{-2}\ (=N\ m^{-2})$	Derived
energy	joule	J	$kg\ m^2\ s^{-2}$	Derived
power	watt	W	$kg\ m^2\ s^{-3}\ (=J\ s^{-1})$	Derived
electric charge	coulomb	C	$A\ s$	Derived
electric potential difference	volt	V	$kg\ m^2 s^{-3} A^{-1}\ (=JA^{-1}s^{-1})$	Derived
electric resistance	ohm	Ω	$kg\ m^2 s^{-3} A^{-2}\ (=V\ A^{-1})$	Derived
electric conductance	siemens	S	$kg^{-1} m^{-2} s^3 A^2\ (=A\ V^{-1}=\Omega^{-1})$	Derived
electrical capacitance	farad	F	$kg^{-1} m^{-2} s^4 A^2\ (=A\ s\ V^{-1})$	Derived
magnetic flux	weber	Wb	$kg\ m^2 s^{-2} A^{-1}\ (=V\ s)$	Derived
inductance	henry	H	$kg\ m^2 s^{-2} A^{-2}\ (=V\ A^{-1})$	Derived
magnetic flux density	tesla	T	$kg\ s^{-2} A^{-1}\ (=V\ s\ m^{-2})$	Derived
luminous flux	lumen	lm	$cd\ sr$	Derived
illumination	lux	lx	$cd\ sr\ m^{-1}$	Derived
frequency	hertz	Hz	s^{-1}	Derived

SI units for other physical quantities can be derived from the seven base units by multiplication or division without introducing numerical factors: this system is *coherent*. This recommendation is often too rigorously interpreted. For example, the exclusive use of SI units (base or derived) is not obligatory when this leads to a unit which is inconvenient in practice (e.g. the farad is too large; microfarad (μF) and picofarad (pF) are commonly used). Any decimal multiples and fractions of SI units may be used, provided they are clearly stated. There are officially recommended names and symbols for powers of ten in frequent use. In spectroscopy, the Ångström is widely used as unit of wavelength (1 Å = 10^{-10} m). It is a decimal fraction of the SI unit of length, the metre. The Ångström has a convenient order of magnitude for the description of optical line spectra and atomic or molecular distances. The nanometer, however, is by one order of magnitude too large, but may be preferred for optical absorption spectroscopy, where relatively wide absorption bands must be described.

16.2.9   Quantitative definitions should, whenever possible, be given by means of equations.

16.2.10   In the nomenclature lists which follow, alternatives are occasionally recognized. If the alternative symbols are separated by a comma, then both are equally usable. Separation by a row of dots signifies that the first-written symbol is preferred.

16.3   TERMS AND SYMBOLS FOR PHYSICAL QUANTITIES IN GENERAL USE

(Extracted from IUPAP 1965)

16.3.1   BASIC PHYSICAL QUANTITIES

length	l
mass	m
time	t
electric current	I
thermodynamic temperature	T
luminous intensity	I_v
amount of substance	n

16.3.2   OTHER PHYSICAL QUANTITIES

space coordinates	x, y, z
breadth (width)	b
height	h
radius	r
area	S, A
volume	V
plane angle	$\alpha, \beta, \gamma, \delta, \theta, \phi$
solid angle	ω, Ω
wavelength	λ
wavenumber ($\sigma = 1/\lambda$)	$\tilde{\nu}, \sigma$ †
period	T
frequency ($f = 1/T$)	ν, f ‡
velocity	v, u
velocity (average)	$\bar{v}, \bar{u}$
speed of light (in a vacuum)	c, c_o

16.3.2 OTHER PHYSICAL QUANTITIES (continued)

relaxation time	τ
radial frequency ($2\pi f$)	ω
pressure	p
force	F
energy	E
power (energy divided by time)	P

Notes: † $\quad \tilde{\nu}$ is exclusively used in molecular spectroscopy

‡ $\quad \nu$ is used in spectroscopy, f in electrotechnics.

16.4 TERMS, SYMBOLS, AND UNITS RELATED TO RADIANT ENERGY

16.4.1 Table 16.4 groups together the nomenclature and symbols used to describe the sources of radiant energy and their activity.

The Table is restricted to a few essential terms. Items 1 and 2 are of a general nature. Items 3,4,5 and 6 refer to the source of the radiation and items 7 and 8 to the receiver or irradiated object. Other radiation quantities with special names, occasionally with special units, occur in the literature, but they are superfluous.

16.4.2 The radiation quantities in Table 16.4, in particular, $\Phi, I, B, J,$ and u, will in general be functions of the wavelength λ (or of the wavenumber $\tilde{\nu}$, the frequency ν, as the chosen variables may be). They may also be functions of other variables, e.g. the temperature or the elements present. Thus it may be necessary to consider the quantity Φ, I, B, etc. within a small region, for example a *bandwidth* of wavelengths. The radiance within such a bandwidth $\Delta\lambda$ would be: $\{dB(\lambda)/d\lambda\}/\Delta\lambda$. For convenience, the differential quotient may be indicated by a subscript, as in $dB/d\lambda = B_\lambda$: B_λ is called the *spectral radiance*. The symbols Φ_λ, I_λ have the corresponding meanings and the adjective spectral is used to distinguish them from the quantities treated in Section 16.4.1.

16.4.3 In spectrochemical analysis, the wavelength λ normally describes a position within a spectrum. In other fields of spectroscopy, wavenumbers $\tilde{\nu}$, frequencies ν, or periods T are used as variables. Therefore it may be useful to list here the relationships of bandwidths expressed in different variables and of the *spectral radiation quantities*, e.g. the spectral radiant flux Φ_λ

$$\lambda\tilde{\nu} = 1 \qquad \lambda\nu = c \qquad \lambda = Tc$$
$$\Delta\lambda = -\Delta\tilde{\nu}/\tilde{\nu}^2 = -c\Delta\nu/\nu^2 = c\Delta T$$
$$\Phi_\lambda = -\tilde{\nu}^2 \Phi_{\tilde{\nu}} = -\nu^2 \Phi_\nu/c = \Phi_T/c$$

16.4.4 The different radiation quantities for the *black body* play an important role in spectroscopy as natural standards because they are determined by constants of nature, temperature and wavelength. They may be distinguished by an additional superscript, the letter b: for example, the *spectral radiance of the black body*, B_λ^b.

16.4.5 If the radiation acts as *light*, affecting the human eye, the radiation may be measured and appreciated in a different way, taking into account not only the physical but

TABLE 16.4

Name	Symbol	Definition	Dimensions	SI unit	Practical unit
1. radiant flux or radiant power	Φ	power in the form of radiation	power	W	W
2. (radiant) energy	Q	$Q = \int_0^t \Phi dt$	energy	J	W s = J
3. radiant intensity	I	$I = \Phi/\omega$	power/solid angle	W sr^{-1}	W sr^{-1}
4. radiance	$B \ldots L$	$B = \Phi/S\omega\cos\epsilon$	power/(area × solid angle)	W sr^{-1} m^{-2}	W sr^{-1} cm^{-2}
5. (radiant) emissivity	J	$J = \Phi/V\omega$	power/(volume × solid angle)	W sr^{-1} m^{-3}	W sr^{-1} cm^{-3}
6. (radiant) energy density	u	$u = Q/V$	energy/volume	J m^{-3}	J cm^{-3}
7. irradiance	E	$E = \Phi/S$	power/area	W m^{-2}	W cm^{-2}
8. (radiant) exposure	H	$H = \int_0^t E dt$	power × time/area	J m^{-2}	W s cm^{-2} = J cm^{-2}

Reference to entries 3 to 8. It must be kept in mind that the basic quantity in this system, the radiant flux Φ, may vary in space and direction. Therefore the area S, the solid angle ω and the volume V in the defining equations 3 to 8 must be small enough to give meaningful local values for the derived quantities.

Reference to entries 4,7, and 8. S refers to the area of the radiating element for the term radiance and to the area of the irradiated element for the terms irradiance and (radiant) exposure, ϵ is the angle between the normal to the surface and the direction of the radiating beam, which is assumed to be confined to a narrow solid angle.

Reference to entry 4. The old symbol B is recommended for radiance in contradiction to IUPAC 1970{*Pure Appl. Chem.*, 21, 1 (1970)}, which only lists L. The letter L should not be used for radiance in atomic emission spectroscopy, since L is also the symbol for the spectrum line, for the orbital quantum number and also for self-inductance.

Reference to entries 7 and 8. The letter E may logically be used as a symbol for radiant exposure. However, E has been internationally adopted for irradiance, while H was chosen as a symbol for exposure. Both quantities are important in photographic photometry, but they are seldom used in the practice of spectrochemical analysis.

Reference to entries 2,5,6, and 8. In these cases the adjective radiant may be omitted, if the meaning of the term is obvious from the context. The adjective radiant should not be dropped from radiant flux and radiant intensity.

Reference to entries 4,5,6,7, and 8. For the practical units it is proposed to use the centimetre as a unit of length. (This is allowed, see paragraph 16.2.8.) The reason for this proposal is obvious: the dimensions of most radiant sources used in spectroscopy are of the order of a centimetre. The use of this unit therefore helps visualization.

also physiological processes. The technique of measurement of light in this context is called *photometry*.

Although in modern spectroscopy visual observations and measurements have become very rare, it seems advisable to give some explanations of photometry. There has been much confusion in the past by inappropriate use of photometric terms and units in the measurement of radiation quantities.

(a) There is a set of luminous quantities which correspond to the radiant quantities of Table 16.4. The same symbols are used. If confusion between luminous and radiant quantities may occur, subscripts should be added: e (energy) for radiant, v(visible) for the luminous quantities. (It has been proposed that in cases where the number of quanta has been determined instead of the energy, the subscript q might be added.)

(b) A relationship between *radiant* and *luminous* quantities has been established by the definition of the *normal eye (standard observer)*. The normal light-adapted eye is characterized by a wavelength dependent function, the (spectral) *luminous efficacy* $K(\lambda)$, whose maximum K_m occurs at 555 nm and has the value K_m = 680 lm/W. The function $K(\lambda)/K_m$ = $V(\lambda)$ is called the (spectral) *luminous efficiency*. The relationship between radiant flux Φ_e, and the luminous flux Φ_v is as follows:

$$\Phi_v = K_m \int_{380 \text{ nm}}^{780 \text{ nm}} V(\lambda) \Phi_{e,\lambda} d\lambda$$

For the exact values of K_m and $V(\lambda)$, the most recent publications from the International Commission on Illumination (CIE *Publication No. 17 (E-1.1) 1970)*) should be consulted.

16.5 TERMS AND SYMBOLS FOR THE DESCRIPTION OF SPECTROGRAPHIC INSTRUMENTS

16.5.1 *Geometrical quantities* - In this section the terms and symbols are described with a *spectrograph* in mind. The modifications necessary to consider other forms of spectral apparatus will be self-evident.

16.5.1.1 In the spectrograph the whole range of spectrum allowed by the design of the instrument is recorded on the photographic plate.

When, in place of the photographic plate, direct means are used to measure intensities in the spectrum, the instrument is described as a *spectrometer*. For example, a thermopile or photoelectric device may be made to scan the spectrum in the focal plane, measuring the intensity at each position.

If the focal plane is obstructed except for one slit, the instrument is called a *monochromator*. The bandwidth of the spectrum which emerges through the exit slit depends *inter alia* upon the widths of the entrance and exit slits. A monochromator is usually provided with means for altering the mean wavelength of the band transmitted, but there are also fixed monochromators which then correspond to a filter.

The *polychromator* is an extension of the monochromator, a number of exit slits being placed in the focal plane, so allowing a number of discrete bands to pass through (to fall, for example, upon a number of photomultipliers).

16.5.1.2 Light enters a spectrograph through a *slit* having a geometric *slit-width s* and *height h*.

16.5.1.3 The light strikes a collimator lens or mirror which has an *aperture stop* of *diameter D*, or the aperture stop may have a rectangular clear cross-sectional *area S* of *width* D_s (parallel to the width of the slit) and *height* D_h. The *effective diameter for a non-circular aperture stop of area S*, (D_{eff}) is given by $(4S/\pi)^{1/2}$ or $(4D_s D_h/\pi)^{1/2}$.

16.5.1.4 The *focal length* of the collimator lens or mirror is f, and the light-gathering power of the lens is measured by the f-number which is f/D_{eff}.

16.5.1.5 The light then encounters a dispersing system (see 16.5.1.9) and after dispersion the light enters an objective lens or mirror, where beams of different wavelengths are focused to form images of the entrance slit at different places along the focal plane.

16.5.1.6 In order to counteract the variation with wavelength of the *focal length*, f', of the objective lens, it may be necessary to tilt the photographic plate, and the *tilt of the plate*, θ, is the angle between the normal to the plate and the direction of the camera axis.

16.5.1.7 The practice of optical designers in distinguishing magnitudes in the image space from similar ones in the object space by using a prime is adopted, hence the f' used in the previous paragraph. In a monochromator, for example, the exit slit has *width s'* and *height h'*.

16.5.1.8 The light during passage through an instrument has a *beam of width W*, and one may distinguish at times W_s and W_h.

16.5.1.9 The dispersing system may be a prism or a series of prisms, and the *length of the total effective prism base* is b. The effective base is the difference between the lengths, measured within the prisms, of the extreme rays near the base and near the apices respectively. Alternatively, the dispersing system may be a diffraction grating, having a *total number of rulings*, N_r. The *number of grating rulings per unit length* is n_r. The rulings may possess a *blaze-angle*, β, which is the angle between the operating facet of the grooves and the overall plane of the grating. This results in the grating enhancing the intensity of wavelengths near to a blaze wavelength λ_β. (The order of the spectrum to which the blaze refers should be stated.) The *order of the spectrum* is preferably denoted by m, although k is accepted.

16.5.2 *Optical quantities*

16.5.2.1 The *refractive index* of a material is n.

16.5.2.2 When a spectral bandwidth is to be indicated in terms of wavelength, the symbol is $\Delta\lambda$. For the width of a spectral line at half-peak intensity - as it appears in a spectrum produced by an instrument - the symbol is $\delta\lambda$. The *width of the line itself* as it would be shown by an instrument of very high resolving power can be written as $\delta_L\lambda$. (This includes the natural width, Doppler and Stark effects, pressure broadening, etc.) The *minimal line width* which can be produced by a spectroscopic instrument for reasons of principle (mostly limited by diffraction) is written $\delta_o\lambda$.

The (theoretical) *resolving power* R_o of a spectroscopic instrument is defined by $R_o = \lambda/\delta_o\lambda$. Very often this resolving power R_o cannot be used or attained for practical reasons (e.g. $\delta_L\lambda > \delta_o\lambda$; optical aberrations and the necessity to use a wide slit). In such cases, the (*practical*) *resolution* is defined by $R = \lambda/\delta\lambda$.

16.5.2.3 *Dispersion* $dz/d\lambda$ is qualified in different ways according to the concepts indicated by z. For example, substitution of z by n gives $dn/d\lambda$ the *dispersion of a material;* by angle ϕ, *angular dispersion;* while substitution of z by the separation x of spectral lines gives the *linear dispersion*.

16.5.2.4 The reciprocal of the last-named quantity is more frequently used, and referred to as the *reciprocal linear dispersion*, $d\lambda/dx$, and it is commonly expressed in Å/mm.

16.5.3 *Quantities related to the transport of radiant energy*

16.5.3.1 Three expressions are used to describe how an optical system transmits, reflects, or absorbs radiation. They are optically composite quantities (for example, the transmission factor of a monochromator), so to indicate this they are called factors. Φ_o is the radiant flux entering the system. The respective terms are:

$$\begin{aligned}
\textit{transmission factor} \quad & \tau = \Phi_t/\Phi_o \\
\textit{reflection factor} \quad & \rho = \Phi_r/\Phi_o \quad \text{and} \\
\textit{absorption factor} \quad & \alpha = \Phi_a/\Phi_o
\end{aligned}$$

16.5.3.2 A little-known but useful quantity is the *optical conductance, G,* that describes the geometrical restriction of the radiant flux through an instrument (or optical system) by the apertures and the distances separating them. This quantity is discuessed in detail in the Appendix.

16.6 TERMS AND SYMBOLS RELATED TO THE ANALYTICAL PROCEDURES

16.6.1 *Qualitative terms concerning the sample*

In analytical spectroscopy, *material* is provided, and from this is taken a *sample* that is submitted for analysis. The material has a certain composition, consisting of an aggregate of *constituents*. If the proportion of one constituent predominates, it is referred to as the *major* constituent. The latter term should be distinguished from the description of the character of the material as a whole, e.g. granite, organic tissues, solutions, which are referred to as the *base*. The element sought or deterninined in the sample is the *analysis element*.

Samples may be transformed into a *solution*. Here we distinguish between the *solvent* (for example, water or a mixture of water and alcohol) and *concomitants*, which include constituents other than the analysis element in the solution.

The term *matrix* refers to the sample, considered as an assemblage of constituents, with all their individual properties. The combined effect that the various constituents of the matrix may exert on the measure of the analysis of the analysis element is referred to as the *matrix effect*.

16.6.2 Quantitative terms concerning the sample

16.6.2.1 The *quantity* of a substance resulting from an analysis is written as q, and the unit in which it is measured must be stated explicitly (e.g. g, mg, µg). The *concentration* is represented by the symbol c, in terms of mass, volume, number, or by any other means, but the units must be expressly stated. For very small concentrations the expression *parts per million* (p.p.m.) by weight may be used without ambiguity for impurities in solids. For trace impurities in solution, the term $\mu g/cm^3$ is less ambiguous than p.p.m. and is strongly recommended. The phrase '*parts per billion*' is ambiguous and its use is discouraged.

16.6.2.2 Normally the *concentration* is expressed relative to the whole sample. The ratio of the concentration of a particular element to that of another element (usually present in high concentration) is called the *concentration ratio* (of X to Y) and is given the symbol c_r. In spectrochemical analysis the denominator is often the concentration of a *reference element*, R, while the numerator is the concentration of the element to be determined, the *analysis element* X (see also Section 16.6.3.2). The *reference element* may be an important component or it may be specially added for the purpose. If *concentration ratios* are used, the two concentrations must be expressed in the same units.

16.6.3 Terms concerning the procedure

16.6.3.1 In emission spectroscopy the physical quantity which is used to characterize and measure the concentration to be determined is referred to as the *intensity*. *Intensity* is one of the most frequently used words in spectroscopy in general and also in spectrochemical analysis. One speaks of the intensity of a spectrum line, of the background, of an absorption band, or of a beam of light, etc. *Intensity* is a generally useful word, guiding one's thoughts in the direction of energy or power, but so far without a definite scientific meaning and therefore an offence to a nomenclature-minded conscience. Therefore the use of the term *intensity* must be made respectable without sacrificing its generality. The aim of every quantitative spectrochemical analysis is to determine a *concentration*, c, or a *quantity*, q, of a substance from the measurements. The particular physical quantity which is derived from the experimental observation (e.g. the pointer-reading of a voltmeter) is related to a radiant quantity, in photographic measurements usually a (radiant) *exposure* H and in photoelectric measurements a *radiant energy* Q. In practice it is often unnecessary to consider what radiant quantity is really measured, or to refer to the particular radiant property of the light source (*radiance* or *radiant intensity*). Then the indefinite expression *intensity* I can be used for the relative strength of the spectral line. Confusion is unlikely to happen with the *radiant intensity* which also has the symbol I; the meaning will be clear from the context. *Intensity of a spectral line or of the background in a spectrum is then a loose relative expression referring to the radiant quantity measured by the receiver. This intensity has unit dimension.* In many cases it is not necessary to state or to determine explicitly to which reference intensity the observed value is referred. The concept of such a reference is always implicit in the use of the term intensity.

The relative nature of *intensity* is explicit when *internal reference lines* in the spectrum are used. Then accidental variations in the physical conditions of the experiment (especially in the light source) are generally without harmful effect, since one measures the intensity I_X of the spectral line of the *analysis element* in relation to the intensity I_R of a line of a suitably selected *reference element* (see also Section 16.8.4.4). Ideally both the analysis element line intensity and the reference line intensity should respond to changes in the experimental conditions in the same way and rate (Gerlach's *homologous lines*).

16.6.3.2 If the *analytical curve* is plotted in logarithmic coordinates for both axes, and concentration ratios are introduced, a straight line is often obtained:

$$\log c_r = \eta \log I_X/I_R + \log c_o$$

which corresponds to the Lomakin-Scheibe equation

$$c_r = c_o (I_X/I_R)^\eta$$

In this equation c_o is often spoken of as the *concentration index*, and from the equation it will be seen that the concentration ratio c_r for which I_X/I_R is unity and consequently $\log I_X/I_R$ is zero. Very often the slope η in the above equation is near to unity.

16.6.3.3 There is normally radiation present at the wavelength of the analysis element and reference element lines that does not arise from the specific electron transitions which produce the line radiation. This extra radiation, which is designated the *background*, may be part of a spectral continuum or of unresolved molecular bands. If the spectrum is photographically recorded this *background* is to be clearly distinguished from the *fog* of the photographic plate, an overall greying of the plate due to the development process, ageing of the plate and similar processes. When it is useful to distinguish quantities concerning a line from similar quantities concerning the background, the subscripts L and U are used respectively (U is taken from the German term *Untergrund* , since B is already used for radiance).

16.6.3.4 In addition to the background, radiation from other nearby spectral lines may perturb the measurement of intensity of the wanted line. Such lines are referred to as *interfering lines*.

16.6.3.5 Another type of unwanted radiation is the light which reaches the receiver in unintended ways. This radiation is identified as *stray light*.

16.7 TERMS AND SYMBOLS RELATED TO FUNDAMENTAL PROCESSES OCCURRING IN LIGHT (EXCITATION) SOURCES

16.7.1 *General rules*

The intensities of spectral lines emitted by a source depend on the fundamental properties of the atoms, molecules, and other particles present, on their relative concentrations, and on the physical conditions prevailing at the source. The system of terms and symbols must therefore be able to provide a simple means of distinguishing these properties, concentrations and conditions, when and to whatever extent it is necessary to do so. To do this with simple means, terms and symbols are employed which are already accepted by IUPAP or by IUPAC.

When, and only when, these symbols refer to different species of particles present at the same time, the correlation to the species may be given by adding in parentheses the symbol of the particle. For example, the mass number of magnesium would be $A(\text{Mg})$.

The electric charge of the species may be indicated at the same time by the superscripts

$$0, +, 2+, \ldots$$

for the neutral, singly-ionized, doubly-ionized .... species (correspondingly for negative ions, substituting the negative sign). According to general usage, accepted by IUPAC and IUPAP, symbols relating to the electron have e as subscript, e.g. N_e represents the number of electrons, P_e the electron pressure. The symbol e by itself represents the electron. (In the equation of equilibrium, where the symbol for the electron may occur as well as the base of natural logarithms, the conflict of symbols may be avoided by the modern usage of 'exp' for the exponential function.)

16.7.2 *Physical constants and properties of particles*

Avogadro constant	N_A
Boltzmann constant	k
Planck constant	h
Gas constant (molar)	R (should be distinguished from the Rydberg constant, R_∞)
Atomic mass	m
Atomic weight	M
Atomic mass of species X	$m(X), m_X$
Atomic weight of species X	$M(X), M_X$
Mass number	A
Atomic number	Z
Mass number of species X	$A(X), A_X$
Atomic number of species X	$Z(X), Z_X$
Elementary charge	e
Electron mass	m_e

16.7.3 *Terms, symbols, and units for measurable quantities*

Number of particles	N
Number of particles of species X	$N(X), N_X$
Number density of particles per unit volume	n
Number density of particles in state q	n_q
Number density of element as free atom	n_{at}
Number density of element as free ion	n_{ion}
Number density of electrons	n_e
Total number density of element in different forms (atom, ion, molecule) in the gaseous state	n_t
Number density of ground state species X	$[X_o]$
n_{at} for atoms of X	$[X], n_{at}(X)$
n_{ion} for ions of X^+	$[X^+], n_{ion}(X)$

n_t for element X	$[X]_t$, $n_t(X)$
Number density of excited species X*	$[X^*]$
Total gas pressure	p, p_t
Partial pressure of species X	$p(X)$, p_X
Excitation energy	E_{exc}
Excitation energy of state of species X	$E_{qK}(X)$, $(E_q)_X$
Excitation potential	V_{exc}
Ionization potential	V_{ion}, V_i
Ionization energy	E_{ion}, E_i
Dissociation energy (minimum energy required to dissociate one molecule or one mole of XY at $T=$ zero K in perfect gas state)	D_o, D_{XY}
Kinetic energy of particle	E_{kin}, E_k
Statistical weight of state	g_q
Statistical weight of ground state	g_o
Statistical weight of state q of species X	$g_q(X)$, $(g_q)_X$
Partition function	Z, Q
Partition function of species X	$Z(X)$, Z_X
Transition probability for the spontaneous transition from state q to state p	A_{qp}, (B_{pq} for the reverse absorption transition)
Degree of dissociation of the species MX, $([M]/[M]+[MX])$	β_d, α
Degree of ionization of species M $([M^+]/[M^o]+[M^+])$	β_i, γ
Dissociation constant of M in equilibrium at T, $([M][X]/[X])$	$K_d(T)$
Ionization constant of M in equilibrium at T, $([M^+][e^-]/[M^o])$	K_i, $K_i(T)$
Frequency of spectral line emitted due to transition $q \to p$	ν_{qp}
'Intensity' of spectral line emitted by electron transitions from states $q \to p$	I_{qp}
Thermodynamic temperature	T

16.7.4 *Conversion factors*

In considering the excitation of particles in a light source, certain values of energy play important roles. These require extensive explanation because the colloquial usage of nomenclature is untidy. To bring a particle into an excited state (or to a higher energy level) a certain amount of energy must be provided. This amount of energy is called the 'excitation energy' and is denoted by E_{exc}. The appropriate SI unit is the joule. To facilitate the calculation of wavelength, the energy levels of particles are often given as (*spectroscopic*) *terms*, T, with unit cm^{-1}. The corresponding energy in joules may be obtained by multiplying the numerical value (in cm^{-1}) by $10^2 hc$ (in $K_q\ cm^3\ s^{-2}$), that is, by a factor close to 2×10^{-23}. Since this factor is constant, the energy may be represented by the corresponding term-values in cm^{-1}, although the dimensions are incorrect.

A similar situation arises through electron-impact experiments. The kinetic energy acquired by an electron in an electric field is given by the product of the electric charge and the difference of potential through which the electron moves.

Since the charge is constant, the kinetic energy of the electron may be represented by giving the value of the potential difference applied in volts. Referring to excitation processes, this potential difference is called the *excitation potential*, V_{exc}, and is measured in volts. For the energy, a special unit has been devised which is numerically equal to the potential difference, called the electron volt† with the symbol eV. As a unit it is incoherent with SI units:

$$\text{one eV equals } 1.6 \times 10^{-19} \text{ J } \ddagger$$

There is a special case of excitation energy just sufficient to free an electron from the particle: this is called the *ionization energy*, E_i.

16.7.5 *Electrical terms*

Quantity of electricity	Q
Potential, potential difference	V
Tension	U, e.g. $U = IR$
Electric current	I
Capacitance	C
Resistance	R
Self-inductance	L
Mutual inductance	M, L_{12}
Reactance	X
Coupling coefficient	k; $k = M/(L_1 L_2)^{\frac{1}{2}}$
Electric field strength	E
Current density If the vector character of E and j is to be brought out, these letters should be printed in heavy type	j
Power	P

16.7.6 *Special terms*

16.7.6.1 Spectral lines. This term originates from the appearance of an atomic spectrum observed with a spectroscope having a high resolution. The individual lines are monochromatic images of the entrance slit. In atomic spectroscopy, this origin has been obscured and the term is now employed to describe a very narrow band of frequencies of electromagnetic radiation resulting from electron transitions in atoms. If the atom has its

---

† The use of the electron volt as an energy unit is discouraged by IUPAC although excitation potentials given in volts fit into the International System of Units.

‡ Slightly rounded.

complete complement of electrons, the transition results in an *atomic line*. This type of line is indicated by the element symbol, followed by I; if one or more electrons are missing, the result is an *ionic line*. To distinguish between the different states of ionization, the type of line is indicated by the element symbol, followed by II, III, IV, etc. (These numbers should be in Roman small capitals.) Examples are: Na I, Mg II, and Al III. The terms 'arc line' and 'spark line' do not distinguish them in the way intended and they should not be used.

16.7.6.2 *Self-absorption* occurs in emission sources of finite thickness when radiant energy quanta emitted by atoms (or molecules) are absorbed by atoms of the same kind present in the same source. The absorbed energy is usually dissipated by collisional transfer of energy, or through emission of radiant energy of the same or other frequencies. In consequence, the observed radiant intensity of a spectral line (or band component) emitted by a source may be less than the radiant intensity would be from an optically thin source having the same number of emitting atoms. Self-absorption may occur in all emitting sources to some degree, whether they are homogeneous or not.

16.7.6.3 *Self-reversal* describes the effect of self-absorption on the shape of the spectral line emitted in light sources that are inhomogeneous in temperature or excitation conditions in the direction of observation. Self-reversal is manifested as a decrease in intensity at the wavelength centre of the line. In extreme cases, the intensity at the wavelength centre may become so weak that practically only the wings remain, giving the appearance of two fuzzy lines.

16.7.6.4 *Intensity versus time curve*. In spectrochemical analysis the study of the variation of the intensity of a spectral line with time is very important. When the aim is to study the volatilization and excitation of a sample for spectrochemical analysis, especially to choose the optimal exposure, the plot of intensity versus time is called the *intensity time curve*. The duration of such a study may be of the order of from tens of seconds to several minutes. The time range selected for the analytical exposure is the *exposure time*. If an initial portion of the excitation cycle is rejected before the analytical exposure is made, these time intervals are designated as *pre-arc* or *pre-spark* periods, or other appropriate terms depending on the light source employed. *Time-resolved spectroscopy* refers to very short exposure periods, e.g. microseconds to milliseconds, employed when physical conditions in the light source are studied.

16.7.7 *Classification of additives*

16.7.7.1 *Additives* are substances added to samples for various purposes. There are many individual names for such substances, which have arisen historically rather than systematically. The following scheme is an attempt to frame a system in such a way that the name indicates the way the additive operates.

16.7.7.2 *Spectrochemical buffers* are added to samples and reference samples with the intention of making the measure of the analytical element less sensitive to changes in concentration of an interferent.

16.7.7.3 A *(spectrochemical) diluent* is a substance added to the sample mainly to increase its bulk for ease of handling. This addition may bring other benefits such as the suppression of undesirable effects due to the previous composition of the material (see Section 16.6.1 *matrix effects*).

16.7.7.4 Material added to a sample to increase its volatilization, or that of some component of it, is a *volatilizer*, while, if it is done for the opposite reason, it is called a *devolatilizer*. Examples of volatilizers are: AlF_3, or NaF, used with uranium, boron or silicon: also all chlorides. A typical devolatilizer is carbon, which gives rise to refractory carbides when used in the analysis of tungsten or boron.

16.7.7.5 A *spectrochemical carrier* is an additive which gives rise to a gas which can help to transport the vapour of the sample material into the excitation region of the source, e.g. carbon in an air atmosphere when carbon dioxide is formed.

16.8 PHOTOGRAPHIC INTENSITY MEASUREMENTS
(PHOTOGRAPHIC PHOTOMETRY)

16.8.1 *Introduction*

Photographic intensity measurements play an important part in spectroscopy and spectro-chemistry. This is due to the enormous information capacity of a photographic emulsion, which produces an image of an extended spectrum in one exposure, showing not only the analysis line under consideration but many other features as well.

The type of such measurements required in spectrochemical analysis is relatively simple; line intensities are to be compared within one spectrum (or a few of them) where the physical conditions are known and can be kept constant. This makes calibration easy and allows a straightforward approach which will be outlined in this section.

16.8.2 *Outline of the measuring procedure*

16.8.2.1 The aim of photographic photometry as used in spectrochemical analysis is the measurement of the comparative intensities of spectrum lines. (Symbols: I and in the case of logarithmic presentation $Y = \log I$.)

16.8.2.2 The physical quantity affecting the emulsion (plate, film) while it is exposed is the *radiant exposure* H. In general practice it is assumed that H is proportional to the intensity I, so usually there is no need to determine H explicitly. This step is tacitly included in the procedure for calibrating an emulsion.

16.8.2.3 The exposure causes a separation of silver in the developed emulsion. A measure of this effect is obtained by measuring the (photographic) *transmittance* T_p at the appropriate place on the plate. The optical instrument used to measure the transmittance T_p is commonly called a *microphotometer*. This name is slightly inappropriate, but it is unambiguous and so widespread that it can be accepted. On the other hand the expression 'densitometer' should not be used for such an instrument.

16.8.2.4 The transmittance T_p is an *auxiliary* and *intermediate* quantity without independent meaning. This is indicated by the subscript p, referring to the photographic plate. The particular numerical value of T_p depends not only on the exposure, the properties of the emulsion and the development, but also on the measuring arrangement (for example on the angular aperture of the light beam in the microphotometer). There is no need to define general standard conditions for measuring T_p. It is only necessary to keep the measuring conditions constant within a series of measurements which are to be compared or correlated with each other.

16.8.3 *Mathematical treatment of measured values for T_p (transformations)*

16.8.3.1 The exposure is generally related in a complicated way to the resulting photographic transmittance T_p, which is measured, so making further calculations inconvenient. It is therefore advisable to transform T_p into a more tractable *photographic parameter P*. Ideally such a transformation should give a linear relationship between the parameter P and the logarithm of the exposure H (or instead, of the line intensity I). But no mathematical transformation — performed only for convenience — can ever improve poor photometric measurements.

16.8.3.2 The relationship between the logarithm of the intensity I and the photographic parameter P is given by the *emulsion calibration function* or in graphical representation by the *emulsion calibration curve*.

16.8.3.3 If, by the selection of a suitable transformation from T_p to a parameter P, the resulting emulsion calibration curve is made straight, the equation of such a straight emulsion calibration curve is

$$P = \gamma_p \log I/I_o$$

or

$$P = \gamma_p (Y - Y_o) \quad \text{with} \quad Y = \log I$$

Here γ_p is the *slope* of the emulsion calibration curve and the index P can be retained if it is clear for what parameter it holds.

I_o and Y_o respectively correspond to the value of the exposure for which $P = 0$.

16.8.3.4 One special class of transformations $T_p \rightarrow P$ leading to (nearly) straight emulsion curves is produced by generalizing the *Baker-Sampson-Seidel transformation* (see also Section 16.8.3.6).

$$P = \kappa \log (1 - T_p) - \log T_p$$

By suitable choice of the *transformation constant* κ, this function is adaptable to many types of emulsions. The numerical values of the two logarithmic functions may be taken from tables or graphs. From this it can be seen that the calibration curve of a photographic emulsion can be characterized by three numbers, e.g. κ, γ_p, Y_o (or I_o), all dependent on wavelength. This must essentially be true for all calibration curves having the same degree of straightness, although, by a special fixation of the transformation from $T_p \rightarrow P$, one of these constants may disappear from the equations, e.g. if γ_p is made unity.

16.8.3.5 The establishment of a suitable transformation can be made by different methods, for instance, by graphical methods resulting in the distortion of the coordinate network or by combining mathematical functions which are known to lead to a linear relationship.

The selection of a suitable transformation is a practical and not a theoretical matter and is irrelevant for the intensity ratios to be determined — if the procedure chosen is correctly performed. Therefore it is proposed to use the symbol P for any photographic parameter leading to a (nearly) straight 'emulsion calibration curve'. If necessary the type of transformation can be indicated in the context. Only when the relative merits of different transformations are discussed, different suitable symbols may be used (e.g. the historic ones from the literature or subscripts).

Unfortunately and most confusingly many different symbols have been used for photographic parameters and for the respective transformations giving a linearization of the emulsion calibration curve, e.g. S, D, W, P, T, L, l, K, Λ, etc., hence the proposal to use only the one symbol P.

16.8.3.6 *Historical note*

The transformation that first served to linearize the emulsion calibration curve, although not in the lower (low intensity) range, is the simple logarithmic transformation which turns the transmittance T_p into the *blackening S*, viz.

$$S = - \log T_p$$

In spite of the partial non-linearity just referred to, this parameter, blackening S, is still widely used (Hurter and Driffield, H and D-curve). The transformation obviously belongs to the class of transformations considered in Section 16.8.3.4 , with the transformation constant $\kappa = 0$. In English the *blackening* is often spoken of as the 'optical density' (D). But this may lead to misunderstanding, especially when 'optical' is omitted. In spectrochemical analysis, the use of the term 'optical density' is therefore discouraged as well as that of the expression 'H and D-curve', for the emulsion calibration curve.

The transformation that first helped to straighten the lower part of the curve was used by two astronomers, Baker and Sampson, in 1924. It was

$$P = \log (1/T_p - 1) = \log (1 - T_p) - \log T_p$$

This is the same equation as in Section 16.8.3.4 if $\kappa = 1$. The same transformation was rediscovered by Seidel in 1936 and subsequently was used extensively in spectrochemical analysis. Later, various transformations have been proposed and investigated (see Section 16.8.3.5).

16.8.4 *Practical calibration of a photographic emulsion*

16.8.4.1 The practical calibration of a photographic emulsion is based on photometric measurements of photographic intensity marks (subscript m) produced by intensities of known ratios. The intensity marks are made with an *intensity calibrating device*, for example, a step filter with a known *transmission ratio* $\tau_m = I_{m,1}/I_{m,2}$. Or one may resort to the use of a pair of spectrum lines of which the intensity ratio is known under stipulated conditions.

16.8.4.2 The calibration procedure is simple if the intensity ratio of two spectrum lines, close together in wavelength, is to be determined (quasi-monochromatic photometry) and if a photographic parameter P, yielding a linear emulsion calibration curve, is used. In this case all that need be measured is the slope γ_p. The values of P_1 and P_2 for the two intensity marks are measured; then the slope of the emulsion calibration curve is

$$\gamma_p = \Delta P_m / \Delta P_m$$

where

$$\Delta P_m = P_1 - P_2$$

and

$$\Delta Y_m = \log I_{m,1} - \log I_{m,2} = \log \tau_m$$

16.8.4.3 Sometimes for convenience — especially with non-intermittent sources — a *step-sector* is used as an intensity calibrating device. In this case the exposure $H = Et$ is varied by variation of t and not by variation of the irradiance E as before. The measured values for ΔP_m in this case may differ markedly from those obtained for the same ratio by variation of the irradiance E.

16.8.4.4 If the wavelengths λ_a and λ_b of the two spectrum lines to be compared differ so much that the emulsion calibration curves are different (heterochromatic photometry), then it may be necessary to use two different photographic parameters P_a and P_b produced by different transformation constants κ, in order to get the two emulsion calibration curves straight. The slopes γ_a and γ_b of these two curves can be determined as before if the transmission ratios τ_a and τ_b of the intensity calibrating device for the two wavelengths are known. In addition to this an *intensity bridge* from λ_a to λ_b is necessary in order to connect the two emulsion calibration curves and so afford the calculation of the difference in response $Y_{a,0} - Y_{b,0}$ of the emulsion at the two wavelengths. An *intensity bridge ratio*

$$\log I_a/I_b = \Delta Y_{a,b}$$

can be derived from a spectrum with known spectral distribution of intensities. Examples are: the spectrum of a standard d.c. graphite arc or of a tungsten ribbon lamp, etc. This idea of an intensity bridge is usually implicit in the term 'external standard'. But the use of this term is discouraged since it is not clearly defined and might wrongly be considered as the opposite of 'internal standard', a term that in future should be replaced by *reference element* or *reference intensity*.

APPENDIX
APPLICATION OF THE CONCEPT OF OPTICAL CONDUCTANCE

16.A.1 *General definition*

Optical conductance, G, is defined by the basic equation

$$\Phi = BG\tau$$

or in words: Flux = Radiance x Optical conductance x Transmission factor, where G embraces the geometric factors. B and Φ are appropriately specified.

16.A.2.1 *Passage of flux between elements of area* of source and sink (e.g. a receiver). dS_1 is an element of area of the source and dS_2 of the sink. a_{12} is the intervening distance, and the normal to dS_1 makes an angle α_1 with the direction from source to sink, and α_2 applies similarly to dS_2. Then, with dS_1 and $dS_2 \ll a_{12}^2$

$$d\Phi = B(\alpha_1) \cos \alpha_1 \, dS_1 \cos \alpha_2 \, dS_2 / a_{12}^2$$

$$= B(\alpha_1) \, dG_{12}$$

16.A.2.2 *Integration.* If in the above dS_1 is an element of an area S_1, and dS_2 of S_2, the total flux from S_1 to S_2 is

$$\Phi = \int_{S_1} \int_{S_2} \dot{B}(x,y,z) \cos \alpha_1 \cos \alpha_2 \, dS_1 \, dS_2 / a_{12}^2$$

If the radiance may be taken as constant over the whole surface S_1, or if it may be replaced adequately by an average value, the radiance may be taken outside the integral, which then constitutes G.

16.A.3 *Generalization to allow for refractive index*

If the medium between S_1 and S_2 has the refractive index n, the optical conductance G_n is defined by

$$G_n = n^2 G_{(n=1)}$$

16.A.4 The practical importance of the concept of optical conductance will be realized if one remembers that the radiance observed within all subsequent apertures of an optical arrangement is invariant — except for losses by reflection, absorption, or scattering which are taken into account by the transmission factor τ.

Therefore, in a correctly designed optical arrangement, consisting of a sequence of parts or instruments (e.g. source - absorption cell - monochromator - receiver), the invariance of the calculated optical conductance between any two apertures in sequence also presupposes that all apertures are fully illuminated. (Of course, some apertures may be larger than necessary; in these cases, only the part filled with radiation must be considered.)

In consequence, the optical conductance of some part in the arrangement that cannot be enlarged for practical or technical reasons determines the effective optical conductance for all other parts and of the whole arrangement. (For example, such a limit for the effective optical conductance may be imposed by the dimensions of an absorption cell (see formula 16.A.5.1) or those of a monochromator (16.A.5.6), a photometer device, or of an interference filter (16.A.5.2).) This fact greatly simplifies basic decisions, the choice of parts or instruments and rough calculations of the available flux.

16.A.5 *Some examples of the use of G of interest in spectroscopy*

16.A.5.1 *Small areas* S_1, S_2: the line joining their centres having length a_{12}, constituting their common normal and S_1 and $S_2 \ll a_{12}^2$

$$G = n^2 S_1 S_2 \, a_{12}^2$$

16.A.5.2 *Radiation through a cone* of vertical semi-angle u, from area S.

$$G = n^2 S \pi \sin^2 u$$

This links the optical conductance treatment with the Helmholtz theorem and the sine condition. Note that $n \sin u$ is the *numerical aperture*.

16.A.5.3 *Sphere of radius* r, radiating into space

$$G = 4\pi^2 n^2 r^2$$

This allows for the approximate treatment of arcs and sparks as small spheres.

16.A.5.4 *Cylinder of radius r and height h* radiating into space

$$G = 2\pi^2 \cdot 2r(h + r)$$

16.A.5.5 *Two surfaces S_1 and S_2* being parts of the inside surface of a sphere of radius r

$$G = S_1 S_2 / 4r^2$$

This shows that G is independent of the relative positions of S_1 and S_2, which provides the basis for (Ulbricht's) integrating sphere.

16.A.5.6 *Monochromator*
16.A.5.6.1

$$G = shD_s D_h / f^2$$

Since the resolving power R_o is $R_o = D_s \, d\phi/d\lambda = (dx/d\lambda) D_s / f$

$$G = (hD_h/f) R_o s \, d\lambda/dx$$

$$= (hD_h/f) R_o \Delta\lambda$$

where $\Delta\lambda$ is the range of wavelengths corresponding to the width of the entrance slit (for the other symbols see Section 16.5).

16.A.5.6.2
If the slit width is measured in units of the half width of the diffraction image, $s_o = \lambda f/D_s$, and similarly if the slit height is measured in the unit $h_o = \lambda f/D_h$, with $\hat{s} = s/s_o$ and $\hat{h} = h/h_o$, then the expression for G becomes very simple,

$$G = \lambda^2 \hat{s}\hat{h}$$

16.A.5.6.3
The value of G in Section 16.A.5.6.1 (with $\Delta\lambda = 1$) may serve as a figure of merit for comparing monochromators, only if the assumption is made that there are no optical imperfections.

16.A.6 *Flux through a monochromator* from a continuous source of radiance, characterized by its *spectral radiance*, B_λ (i.e. $dB(\lambda)/d\lambda$),

$$\Phi = B_\lambda \Delta\lambda \tau (hD_h/f) R_o \Delta\lambda$$

$$\Phi = B_\lambda (\Delta\lambda)^2 \tau R_o (hD_h/f)$$

Therefore the square of the bandwidth is important. The monochromator itself is characterized in this equation by the product

$$Z_M = \tau R_o (hD_h/f)$$

which obviously is a figure of merit for the transport of energy. This equation may also be written (from 16.A.5.6.1) as

$$Z_M = \tau G / \Delta\lambda$$

which is the optical conductance per unit bandwidth.

17. NOMENCLATURE, SYMBOLS, UNITS AND THEIR USAGE IN SPECTROCHEMICAL ANALYSIS—II. DATA INTERPRETATION*

17.1 INTRODUCTION

The present document is concerned with the nomenclature and symbols related to data interpretation common to all of the specific fields of spectrochemical analysis. One purpose is to provide the simplest format for describing quantitative results with carefully defined limits of accuracy. Another purpose is to standardize terms and symbols which will allow easy exchange of information and analytical procedures.

17.2 GENERAL CONCEPTS

17.2.1 MEASURE OF CONCENTRATION AND QUANTITY

In all fields of spectrochemical analysis, a quantitative *measure*, x, of some characteristic feature (e.g. a spectral band, edge, etc.) of the analyte, i.e. the analysis element, is observed. The *concentration*, c, or the *quantity*, q, of a substance contained in a sample must be derived from the observed measure. Random and systematic uncertainties in the value of x itself and in its relationship to c or q determine the precision and accuracy of the analysis.

17.2.2 SENSITIVITY

A method is said to be sensitive if a small change in concentration, c, or quantity, q, causes a large change in the measure, x, that is, when the derivative dx/dc or dx/dq is large. The *sensitivity*, S_i, for element i is defined as the slope of the analytical curve (see Section 17.3). S_i may vary with the magnitude of c_i or q_i; at low values of c_i or q_i, S_i is usually constant. S_i may also be a function of the c or q of other analytes present in the sample.

Note: The term sensitivity has occasionally been misused in atomic absorption spectroscopy to denote the concentration required to cause 1% absorption. The term sensitivity has also been misused to denote the limit of detection. All such misuses are discouraged because they only make communication between analysts more difficult (see Part III, Section 18.4.2).

17.2.3 STANDARD DEVIATION

If the same measurement is repeated n times, the values observed for x will not be exactly the same each time. A useful term describing the random variation in x is the *standard deviation*, s.

---

* Based upon the approved Recommendations published in *Pure and Applied Chemistry*, Vol. 45, No. 2 (1976), pp. 99 - 103.

The value of s is given by the expression

$$s = \left[\sum_{j=1}^{n} (x_j - \bar{x})^2 / (n - 1) \right]^{1/2} \qquad 17.2.3$$

where x_j is an individual measurement. In a precise sense equation 17.2.3 can give the correct value, σ, of the standard deviation of the whole population only if n is an infinitely large number. When n is a small number, say 10, the symbol s should be used instead of σ to indicate that the value of the standard deviation is only an estimate obtained from a small number of measurements.

Notes: (i) The number of observations on which the reported standard and/or relative standard deviations were calculated should always be given. Furthermore, the procedures followed in obtaining the standard deviation should be described.

(ii) By analogy, it is understood that the general equation 17.2.3 for the calculation of σ, s, or s_r can be applied to determined concentrations, c, or to quantities, q. The values of σ, s, or s_r calculated for c or q generally will not agree with the corresponding value for the analyte measures.

17.2.4 RELATIVE STANDARD DEVIATION

Relative standard deviation, s_r, is simply s divided by $\bar{x}$. It is preferably expressed as a decimal fraction but may be expressed in per cent (by multiplying by 100) in those cases where possible confusion with per cent concentration does not arise.

17.2.5 VARIANCE

Several factors contribute to the random uncertainty in any measurement or determination, e.g. random variations in the number of photons emitted or absorbed, variations in setting the instrument at the desired position, and contamination by reagents. Each of these factors contributes to the standard deviation of the final result according to the rules of *variance*. The total variance is given by the expression

$$s_T^2 = s_1^2 + s_2^2 + s_3^2 + \ldots s_m^2 \qquad 17.2.5$$

where the subscripts refer to statistically independent factors contributing to the uncertainty.

In particular, background or blank corrections must be made for most spectrochemical procedures and the *background* s_b or *blank* s_{bl} standard deviations are some of the terms contributing to s_T. (The limitations introduced by s_b are more fully treated in Section 17.4.1.)

17.2.6 PRECISION

The random uncertainty in the value for the measure, x, or the corresponding uncertainty in the estimate of the concentration, c, or quantity, q, is represented by precision, which is conveniently expressed by the term standard deviation or relative standard deviation as discussed in the previous Sections 17.2.3 or 17.2.4. For multicomponent systems, the precision in c_i, or q_i, for the element i may depend not only on the precision of x_i but also on the precision of x_i for each of the other elements present.

17.2.7 ACCURACY

Accuracy relates to the agreement between the measured concentration and the '*true value*'. The principal limitations on accuracy are (a) random errors (see Section 17.2.6) ; (b) systematic errors due to *bias* in a given analytical procedure: bias represents the positive or negative deviation of the mean analytical result from the known or assumed true value; and (c) in multicomponent systems of elements, the treatment of interelement effects may involve some degree of approximation that leads to reproducible but incorrect estimates of concentration.

17.3 ANALYTICAL FUNCTIONS AND CURVES

17.3.1 SYSTEMS WITHOUT INTERELEMENT EFFECTS

In general, the relation of the measure x to the concentration c or quantity q is called the *analytical function*. A graphical plot of the analytical function, whatever the coordinate axes used, is called the *analytical curve*.

For one-component systems or multicomponent systems for which interelement effects can be neglected, the measure x of element i can be expressed as a function of the concentration c, or quantity q, i.e. $x_i = g_i(c_i)$ or $x_i = g_i(q_i)$. These functions are called the *analytical calibration functions*; the graphs corresponding to these functions are called the *analytical calibration curves* and are determined by observations on reference samples of known concentrations.

The *analytical evaluation functions*, $c_i = f_i(x_i)$ or $q_i = f_i(x_i)$, are often used; their corresponding graphs are called *analytical evaluation curves*. These curves are derived from analytical calibration curves by interchanging the x and c or q axes. The distinction between analytical evaluation and analytical calibration functions may at first sight seem superfluous. This distinction may be trivial in the case of analysis for one-component systems, but assumes importance for multicomponent systems when the measures for the individual components are interdependent because of various interelement effects.

17.3.2 SYSTEMS WITH INTERELEMENT EFFECTS

The measure x_i for the element i may depend not only on the concentration c_i (or quantity q_i) but also on the concentrations or quantities of other elements present. The analytical calibration functions then take the form

$$x_i = g_i(c_1, c_2, c_3, \ldots c_n) \qquad 17.3.3$$

and the analytical evaluation functions take the form

$$c_i = f_i(x_1, x_2, x_3, \ldots x_n) \qquad 17.3.4$$

These functional relationships can be expressed in various approximate forms. In the simplest approximation, the effect of element j on element i may be expressed as a constant multiplier α_{ij} to give a set of linear equations

$$c_i = \sum_j \alpha_{ij} x_j \qquad 17.3.5$$

This approximation may be valid only over a small range of variations of the values of c.

In special cases, non-linear analytical functions may be linearized, in good approximation, by introducing new sets of variables which are suitable functions of c_i or x_i.

17.4 TERMS RELATING TO SMALL CONCENTRATIONS

17.4.1 LIMIT OF DETECTION

The *limit of detection*, expressed as the concentration, c_L, or the quantity, q_L, is derived from the smallest measure, x_L, that can be detected with reasonable certainty for a given analytical procedure. The value of x_L is given by the equation

$$x_L = \bar{x}_{bl} + k s_{bl} \qquad 17.4.2$$

where $\bar{x}_{bl}$ is the mean of the blank measures and s_{bl} the standard deviation of the blank measures and k is a numerical factor chosen according to the confidence level desired. In this context, blank measures x_{bl} refer to the measures observed on a sample that does not intentionally contain the analyte and has essentially the same composition as the material under study. The value of s_{bl} must be determined from the measuring conditions to be used for evaluating x_L and $\bar{x}_{bl}$. The minimum concentration or quantity detectable is, therefore, the concentration or quantity corresponding to

$$c_L = (x_L - \bar{x}_{bl})/S \qquad 17.4.3$$

$$q_L = (x_L - \bar{x}_{bl})/S \qquad 17.4.4$$

where S is assumed to be constant for low values of c or q. The values for $\bar{x}_{bl}$ and s_{bl} cannot usually be determined from theory but must be found experimentally by making a sufficiently large number of measurements, say 20. (When counting statistics are involved, as in X-ray spectroscopy, s_{bl} is often estimated directly from a single measurement of s_b because $x_b \cong N_b$, the number of photons and $s_b \cong \sqrt{N_b}$, if Poisson statistics are followed.)

A value of 3 for k in equation 17.4.2 is strongly recommended; for this value a 99.6% confidence level applies only for a strictly one-sided Gaussian distribution. At low concentrations, non-Gaussian distributions are more likely.

Moreover, the values of $\bar{x}_{bl}$ and s_{bl} are themselves only estimates based on limited numbers of measurements. Therefore, in a practical sense, the $3s_b$ value usually corresponds to a confidence level of about 90%.

17.5 GLOSSARY OF TERMS AND SYMBOLS USED

Term	Symbol
Concentration, of element i	c, c_i
Quantity, of element i	q, q_i
Measure, for element i, average value	$x, x_i, \bar{x}$
Sensitivity	S
Relative standard deviation	s_r
Analytical calibration function	$x = g(c)$ or $x = g(q)$
Analytical evaluation	$c = f(x)$ or $q = f(x)$
Limit of detection	c_L

18. NOMENCLATURE, SYMBOLS, UNITS AND THEIR USAGE IN SPECTROCHEMICAL ANALYSIS—III. ANALYTICAL FLAME SPECTROSCOPY AND ASSOCIATED NON-FLAME PROCEDURES*

18.1 INTRODUCTION

Part III is a sequel to Parts I (*Pure Appl. Chem.*, 30, 653 (1972)) and II (*Pure Appl. Chem.*, 45, (2) 99 - 103 (1976)) of the Nomenclature for Spectrochemical Analysis. Whereas Parts I and II are mainly concerned with some general recommendations, Part III deals specifically with analytical flame spectroscopy and associated procedures.

The purpose of this nomenclature, as well as its adaptation to the documents developed by IUPAC and IUPAP in more general fields of chemistry and physics, has already been explained in the Foreword of Part I. In the development of Part III, the recommendations of IUPAC, IUPAP and the International Commission on Illumination (CIE), and of the foregoing Parts I and II, were taken as the starting point. Deviations from these recommendations occur only in some exceptional cases and mainly concern the choice of a consistent set of symbols within the restricted field of analytical flame spectroscopy. Deviations may also be found in a few cases where previous international documents are not in agreement among themselves. Such deviations are explicitly identified in the *Notes* added to this document.

In those instances of specific usage in flame spectroscopy which are not already established by international conventions, compromises often had to be made between the historically developed nomenclature and terminology and those based on more logical considerations. Some expressions which have found their way into the language of practical analysts, but which could be misleading, have been abandoned. The use of alternative terms for the same item was considered undesirable and, as a rule, only one term has been recommended for each item. Some differences in terminology that developed historically among the practioners of emission and atomic absorption methods have been reconciled.

During the period in which this document was in preparation, several national organizations developed their own nomenclature for atomic absorption spectroscopy. Through mutual deliberations, serious conflicts between national documents of restricted scope and the IUPAC nomenclature were significantly reduced. The simultaneous emergence of these national initiatives clearly proves the general need for a well-defined terminology in this field and calls for international cooperation.

Part III is concerned with the analytical application of flame spectroscopy by emission, absorption, and fluorescence methods. These three branches of analysis have many terms in common and a uniform terminology seems mandatory. In the context of this document, an "ordinary" *flame* may be defined as a continuously flowing gas mixture at atmospheric

* Based on the approved Recommendations published in *Pure and Applied Chemistry*, Vol.45 (1976), pp. 105 - 123.

pressure, which emerges from a burner and is heated by combustion. Although this document is primarily concerned with systems that comprise such ordinary flames, burners, and nebulizers, attention is also paid to similar procedures comprising other sampling, atomizing, and/or exciting devices (see Section 18.3.1.3). These procedures bear some resemblance to ordinary flame spectroscopy with regard to the methods of measuring emission, absorption, or fluorescence signals, as well as to the rather simple nature of the spectra involved and the simplicity of the spectra obtained. Arc and spark spectroscopy involving more elaborate and expensive instrumentation are not discussed in this document.

This document is subdivided into several sections corresponding to the different aspects of analytical flame spectroscopy. A general classification and consistent terminology of the different branches of flame spectroscopy are presented in Table 18.1. For some terms, abbreviations are suggested for practical usage. To comply with the internationally agreed restricted meaning of "photometer" (see Section 16.4.5 in Part I), the term "flame photometry" had to be abandoned although it was realized that many practitioners of flame emission spectroscopy may have difficulty in giving up this long-cherished term. On the other hand, the abandonment of this term removes the illogical juxtaposition of terms such as "emission flame photometry" and "atomic absorption spectroscopy" which are often found in the literature. The more uniform terminology presented in Table 18.1 should help to bridge the gap between absorption and emission methods.

Descriptive terms denoting processes or instrumental components and terms for measurable quantities are presented separately. The descriptive terms and some of the most important quantitative terms are explained in a narrative form. To facilitate reference, all quantitative terms are presented in Tables, together with their symbols and practical units. Notes have been added to the text as well as to the tables in order to provide additional explanation or justification, or to warn against improper usage of terms. These notes may readily be passed over on first reading, as all specifically recommended terms are incorporated in the text or in the tables themselves. Such terms, when first met and defined in the text, are printed in italics for easy recognition.

This nomenclature is not intended to provide a comprehensive set of terms. Definitions and explanations are only given when this is required to make their meaning unambiguous. General chemical or physical concepts and quantities are presented in most cases without further elucidation. To facilitate the use of this document, terms and symbols used in flame spectroscopy which were presented in the foregoing Parts I and II have been recapitulated. This applies in particular to the general quantitative terms listed in Tables 18.2 and 18.5. Important terms such as "concentration", "analytical curve", "sensitivity", etc. are briefly explained in Section 18.4 with explicit reference to Parts I and II for a fuller explanation.

In view of the limited number of letters in the alphabet, symbols have been selected carefully, while making allowance for existing international recommendations as well as for long usage. In some cases, alternative symbols have been added, which allows the user to select a set of symbols that is most consistent in a particular context. Alternative symbols that are recommended without preference are listed after each other and separated by commas. When a particular symbol is preferred, it is listed first while the alternative symbols are separated from it by a row of dots. The printing of symbols and indices should conform to the general international rules recapitulated in Section 16.2.3 of Part I.

TABLE 18.1 Classification of methods and instruments

	Absorption	Emission	Fluorescence
Methods			
General classification	Absorption spectroscopy†	Emission spectroscopy†	Fluorescence spectroscopy†
Instruments			
General classification	Absorption spectrometer‡	Emission spectrometer‡	Fluorescence spectrometer‡
When atomic lines are observed§	Atomic ansorption spectroscopy (AAS) Atomic absorption spectrometer	Atomic emission spectroscopy (AES) Atomic emission spectrometer	Atomic fluorescence spectroscopy (AFS) Atomic fluorescence spectrometer
When a flame is used as a means for vaporization, atomization and/or excitation¶	Flame absorption spectroscopy (FAS) Flame absorption spectrometer	Flame emission spectroscopy (FES) Flame emission spectrometer	Flame fluorescence spectroscopy (FFS) Flame fluorescence spectrometer
When both a flame is used and atomic lines are observed	Flame atomic absorption spectroscopy (FAAS) Flame atomic absorption spectrometer	Flame atomic emission spectroscopy (FAES) Flame atomic emission spectrometer	Flame atomic fluorescence spectroscopy (FAFS) Flame atomic fluorescence spectrometer

Note: The term flame photometry (flame photometer) has been abandoned (see also Part I, Section 16.4.5).

†Spectroscopy may be replaced by the more restrictive term *spectrometry* when quantitative measurements of intensities at one or more wavelengths are performed with a spectrometer (see below).

‡The term *spectrometer* as it is used here implies that quantitative measurements of intensities at one or more wavelengths are performed with a photoelectric detector. Wavelength selection may be accomplished, e.g. with a monochromator or optical filter.

§When molecular species are observed, "molecular" is substituted for "atomic".

¶Alternative, but presently less common, methods of vaporization, atomization and/or excitation are, for example, furnaces, flame-like (electrical) plasmas, and cathodic sputtering tubes (see Section 18.3.13). The appropriate adjective should then replace the term flame.

18.2 TERMS AND SYMBOLS FOR GENERAL QUANTITIES AND CONSTANTS

Table 18.2 lists terms and symbols for some general physical and chemical quantities which are commonly used in analytical spectroscopy. This Table is partly an abstract from, and partly forms a supplement to, Sections 16.3.1 and 16.7.2 of Part I of the Nomenclature in Spectrochemical Analysis.

18.3 TERMS, SYMBOLS, AND UNITS FOR THE DESCRIPTION OF THE ANALYTICAL APPARATUS

The functions of an analytical flame spectromter in general are:
(a) Transformation of the solution to be analysed into a vapour containing free atoms or molecular compounds of the analyte (see Section 18.4) in the flame;
(b) Selection and detection of the optical signal (arising from the analyte vapour) which carries information on the kind and concentration of the analyte;

(c) Amplification and read-out of the electrical signal. Terms for the description of component parts of a flame spectrometer (and similar systems) and of processes occurring therein are discussed below. Terms for the description of processes and properties related in particular to the gaseous state of matter in flames are discussed in Section 18.5.

TABLE 18.2 Terms and symbols for general quantities and constants

Terms	Symbol	Note
Mass	m	
Atomic weight (relative atomic mass) of species X ($A_r = 12$ for ^{12}C)	$(A_r)_X, A_r(X)$	
Atomic mass of species X	$m_X, m(X)$	
Volume	V	
Solid angle	Ω, ω	
Time	t	
Frequency (in optical spectroscopy)	ν	
Frequency (in electrotechnics)	f	
Wavelength	λ	
Wavenumber ($1/\lambda$)	$\sigma, \tilde{\nu}$	
Gas pressure	p	
Total pressure of gas mixture	p_t	
Partial pressure of species X	$p_X, p(X)$	
Number of particles	N	
(Number) density of particles (number per unit volume)	n	
(Number) density of species X	$n_X, n(X), X$	
Thermodynamic (or absolute) temperature	T	The unit for T is the Kelvin (K)
Velocity of light (*in vacuo*)	c	
Gas constant	R	
Avogadro number	N_A	
Boltzmann constant	k	
Planck constant	h	
Elementary charge	e	The symbol for "electron" is e⁻ and should not be printed in italics.

18.3.1 TRANSFORMATION OF SAMPLE INTO VAPOUR

18.3.1.1. Descriptive terms concerning nebulizer-flame systems

18.3.1.1.1 *Nebulization, desolvation, volatilization, and atomization*. With a *pneumatic nebulizer* driven under the action of a compressed gas stream, the solution is *aspirated* from the sample container and *nebulized* into a *mist* or *aerosol* of fine droplets. The term *sprayer* denotes that particular part of a nebulizer where the aspirated liquid is disrupted by the gas-jet into a spray.

By *desolvation*, i.e. evaporation of the solvent from the droplets, this mist is converted into a *dry aerosol* consisting of a suspension of solid or molten particles of the solute.

In the high-temperature environment of the flame, *volatilization* of these particles follows.

In atomic flame spectroscopy, the *atomization*, i.e. the conversion of volatilized analyte into free atoms, should be as complete as possible in order to obtain a maximum signal. Any system which is capable of converting the analyte into atomic vapour is called an *atomizer* (see also Section 18.3.1.3).

18.3.1.1.2 *Nebulizers*. In the *chamber-type nebulizer*, the nebulizing gas-jet stream emerges from the sprayer into a *spray chamber*. In such a chamber, the gas-jet stream is homogeneously mixed with the mist droplets. Some of these droplets may evaporate, coalesce, or deposit on the chamber walls and subsequently drain off as waste.

> Nebulizers can be described as follows:
> According to the source of energy used for nebulization as, for example, *pneumatic* or *ultrasonic nebulizers*.
> According to the way the liquid is taken up, e.g. *suction*, *gravity-fed*, *controlled flow*, and *reflux-nebulizers*.
> According to the relative position of the capillaries for the nebulizing gas and the aspirated liquid, e.g. *angular* and *concentric nebulizers*.

Special devices are the *nebulizer with heated spray*, the *twin nebulizer*, and the *drop generator*.

Note: Nebulizers by themselves should not be called atomizers.

18.3.1.1.3 *Burners and flames*. Flames are produced by means of *burner* to which *fuel* and *oxidant* are supplied, usually in the form of gases. With the *premix burner*, fuel and oxidant are thoroughly mixed inside the burner housing before they leave the burner ports and enter the *primary-combustion* or *inner zone* of the flame. This type of burner usually produces an approximately *laminar* flame, and is commonly combined with a separate unit for nebulizing the sample.

In contrast, a *direct-injection burner* combines the function of nebulizer and burner. Here oxidant and fuel emerge from separate ports and are mixed above the burner orifices through their turbulent motion. The flame produced by such a burner is *turbulent*. Most commonly, the oxidant is also used for aspirating and nebulizing the sample. However, when the fuel is used for this purpose, the term *reversed direct-injection burner* is applied. In each case, the mist droplets enter the flame directly, without passing through a spray chamber.

Note: The term total-consumption burner, which is often used, is not recommended.

Premix burners are distinguished as *Bunsen-*, *Méker-*, *slot-burners* according to whether they have one large hole, a number of small holes, or a slot as outlet port(s) for the gas mixture, respectively. When several parallel slots are present, they are identified as *multislot burners* (e.g. a *three-slot burner*). The small diameter of the holes in the Méker burner or the narrowness of the slot in the slot-burner prevents the unwanted *flash-back* of the flame into the burner housing.

At the edge of the flame where the hot gas comes into contact with the surrounding air, secondary combustion occurs and the *secondary combustion* or *outer zone* is formed. The region of the flame confined by the inner and outer zones, where in many instances the conditions for flame analysis are optimum, is called the *interzonal region*, or, when the combustion zones have the form of a cone, the *interconal zone*.

Sometimes provision is made to screen the observed portion of the flame gases from direct contact with the surrounding air. This may be done either mechanically by placing a tube on the top of the burner around the flame, which produces a zonal separation (*separated flame*) or aerodynamically by surrounding the flame with a sheath of inert gas that emerges from openings at the rim of the burner top (*shielded flame*). Observations can thus be made without disturbances from the secondary-combustion zone.

To promote the atomization of elements that readily form oxides in the vapour phase in the flame, a *fuel-rich flame* is often chosen, where reducing conditions favour the dissociation of the metal oxides.

18.3.1.2 Terms, symbols, and units for measurable quantities relating to nebulizer-flame systems.

An easily measurable quantity is the *rate of liquid consumption* by the nebulizing system, defined as the volume of liquid sample that is consumed per unit of time (symbol: F_l, see Table 18.3.1). In particular, in the common case of a pneumatic nebulizer, the term *rate of liquid aspiration* is more specific.

Often only a fraction of the analyte solution that is aspirated passes through the flame cross-section at the observation height (see Table 18.3.2) in a form that is accessible for spectroscopic observation. There are losses of different kinds that limit this fraction and consequently the sensitivity of the method. Examples of such losses with premix burners are the waste of solution in the spray chamber, the burner, and the tubes between them, because of deposition of mist droplets on the walls. With direct-injection burners, there may be losses of droplets that are ejected from the flame because of the turbulent motion of the gases leaving the burner. Moreover, the residence time of the larger droplets in the flame may be insufficient for complete desolvation (see Section 18.3.1.1.1). Similarly, the particles formed after desolvation may not be completely volatilized (see Section 18.3.1.1.1) at the observation height. Finally, it should be recognized that only part of the vapour produced by the analyte may consist of free atoms.

To describe these losses quantitatively, the following terms are recommended. These terms relate to the amount of analyte aspirated per second and entering the flame per second, or passing through the total horizontal flame cross-section per second at the observation height in different states.

The *efficiency of nebulization*, ε_n, is the ratio of the amount of analyte entering the flame to the amount of analyte aspirated.

Note: The quantity ε_n is not related to the amount of solvent but to the amount of analyte. Its value cannot be determined unambiguously by simply comparing the volume of solution drained per second from the spray chamber with the aspiration rate. Correction must usually be made for the difference in analyte concentration in the drained and aspirated solutions respectively, due to the partial evaporation of solvent from the mist droplets deposited on the walls.

The quantity ε_n is not merely characteristic of the operation of the nebulizer, but of the nebulizer - burner system as a whole.

The *(local) fraction desolvated*, β_s, is the ratio of the amount of analyte passing in the desolvated state (i.e. either as a dry aerosol or as a vapour) to the total amount of analyte passing. When this fraction varies with height, due to progressive evaporation of the aerosol droplets in the flame, it is appropriate to speak of the local fraction desolvated.

Notes: Losses due to incomplete volatilization of the dry aerosol (which depend largely on the nature and concentration of the solute) are not covered by the definition of β_s but by the definition of β_v. The quantity β_v will usually depend on the solute whereas β_s will depend on the solvent.

When this ratio varies markedly with the height over the volume of the flame observed, the amount of analyte found in this volume in the considered state is related to the average value of this ratio over this volume.

The (*local*) *fraction volatilized*, β_v, is the ratio of the total amount of analyte passing in the gaseous state to the total amount of analyte passing in the desolvated state. The gaseous state includes free atoms as well as molecules.

The (*local*) *fraction atomized*, β_a, is the ratio of the amount of analyte passing as free neutral (or ionized) atoms to the total amount of analyte passing in the gaseous state.

Note: This fraction is determined by chemical reactions in the gaseous state. The bond strengths of the molecular compounds which the analyte may form in the flame play an important part as well as the composition and the temperature of the flame.

The overall (*local*) *efficiency of atomization*, ε_a, is defined as the ratio of the amount of analyte that passes through the flame cross-section at the observation height, as free neutral (or ionized) atoms, to the amount of analyte aspirated. Therefore, $\varepsilon_a = \varepsilon_n \beta_s \beta_v \beta_a$. The atomic signal strength obtained for a given solution concentration is proportional to the product $F_l \varepsilon_a$. It is noted that ε_a may depend on F_l.

In Table 18.3.1, the above quantitative terms and some further terms for measurable quantities belonging to this section are listed, together with their symbols and units.

18.3.1.3 Terms concerning special sampling, atomizing, and exciting devices.

18.3.1.3.1 *Special sampling devices for flames*. Samples may be introduced into flames by means other than nebulizers. The samples may be deposited on a *sampling loop*, *sampling boat*, or *sampling cup* made from platinum, tungsten, or other high melting point materials, and subsequently thermally vaporized.

Note: These devices are heated only by the flame and not by an additional source of energy.

In atomic absorption spectroscopy a *long tube device* is sometimes used with a nebulizer-flame system to increase the sensitivity. The increase in sensitivity is achieved by "retaining" the combustion gases with the atomized analyte over an extended path length by means of a tube coaxial with the optical axis. The tube is made from material capable of withstanding the flame temperature. The flame gases enter the tube at one end and leave at the other, or when a T-shaped arrangement is used, the gases enter at the centre and flow toward the two open ends. T-shaped tubes are called *T-tubes*.

Note: The long tube has often been referred to as an absorption cell. However, the term absorption should be reserved for devices closed by optical windows.

18.3.1.3.2 *Electrical flame-like plasmas for atomization and excitation*. *Electrical flame-like plasmas* may be *current-carrying plasmas* or *current-free plasmas*. Plasmas may be formed by an arc discharge in a chamber and transferred through an appropriate opening to form a *plasma jet*. Flame-like plasmas can also be generated by high-frequency fields with or without the use of electrodes. With the *electrodeless plasma*, the electromagnetic field is inductively coupled to the plasma (*inductively coupled plasma*). The *single-electrode plasma* is formed on a metallic tip that is connected to a high-frequency generator.

Notes: Both chamber-type nebulizer systems and nebulizers which spray the aerosol directly into the plasma are used to introduce the sample.

The term "plasma-flame", often used for flame-like plasmas, is not recommended. The term "flame" should be reserved for hot gases which are produced by combustion. Similarly, the term "plasm burner" is discouraged. As a general descriptive term, "plasma torch" is often used.

18.3.1.3.3 Non-flame atomizing devices.

18.3.1.3.3.1 *Resistance-heated devices*. When small amounts of liquids are to be analysed or if the sample is to be atomized directly from the solid state, different types of atomizing devices with electrical resistance heating can be used. The sample can be introduced on, or into, an electrically conductive support made of a material with a high melting point and heated by an electrical current. The device may be specified according to the material and shape, such as a *carbon* or *metal filament, loop, ribbon* or *braid atomizer*.
Atomizing devices using heated carbon or graphite tubes are called *carbon-* or *graphite-tube furnaces*. If a rod is used with a hole drilled perpendicularly to its axis, such furnaces are referred to as *carbon-* or *graphite-rod furnaces*. Cup-shaped atomizers used most often in AFS are the *carbon-* or *graphite-cup atomizers*.

Normally an electrical current flowing through the walls of the support causes the temperature to rise by resistance heating. Other less often used types have a separate resistance wire wound round the walls of the tubes or rods. Another type of resistance-heated graphite-tube furnace is used with a d.c. arc discharge to accelerate the atomizing process. In this case, the sample is not placed directly into the tube but on the tip of an anode electrode and evaporated by the arc into the heated tube.

Note: The terms "graphite cell" or "graphite cuvette" should be reserved for devices with closed ends.

18.3.1.3.3.2 *Hollow-cathode devices*. In AAS, a *hollow-cathode discharge* can also be used as an atomizer when the sample is used either as a cathode or placed in the hollow cathode in a low-pressure discharge chamber. In a *cooled hollow cathode* the cathode cylinder is cooled by water, liquid nitrogen, or other means. Under such conditions samples are atomized by *cathodic sputtering* even at high current densities. In a *hot hollow cathode*, samples are atomized primarily by thermal evaporation.

18.3.1.3.3.3 *Radiation-heated devices*. Solid samples can be evaporated and atomized by radiation sources, such as *pulsed-discharge lamps* and *lasers*. Because a laser beam can be brought to a sharp focal point on the sample surface, a *local analysis* can be made with this device.

TABLE 18.3.1 Transformation of sample into vapour. Terms, symbols, and units for measurable quantities

Terms	Symbol	Practical unit	Note
Rate of liquid consumption	F_l	$cm^3 s^{-1}$	For definition, see text of Section 18.3.1.2. In the usual case of a pneumatic nebulizer F_l is called the rate of liquid aspiration
Efficiency of nebulization	ε_n	1	For definition see the text of Section 18.3.1.2
Fraction desolvated	β_s	1	For definition see the text of Section 18.3.1.2
Fraction volatilized	β_v	1	For definition see the text of Section 18.3.1.2
Fraction atomized	β_a	1	For definition see the text of Section 18.3.1.2
Efficiency of atomization	ε_a	1	For definition see the text of Section 18.3.1.2
Flame temperature	T_f	K	When the temperature varies locally in the flame, it is more appropriate to speak of the local flame temperature
Travel time (time needed for substance to be carried from base of flame to the observation volume)	t_{tv}	s	
Transit time (time needed for substance to pass through the observation volume)	t_{ts}	s	
(Vertical) rise velocity of the flame gas	v_r	$cm\ s^{-1}$	
Burning velocity (of flame front)	v_b	$cm\ s^{-1}$	
Flow-rate of unburnt gas mixture	F_u	$cm^3 s^{-1}$	Measured at atmospheric pressure and room temperature
Flow rate of species X, e.g. air, O_2, etc.	F_X	$cm^3 s^{-1}$	Measured at atmospheric pressure and room temperature

18.3.2 LIGHT SOURCES IN ATOMIC ABSORPTION AND ATOMIC FLUORESCENCE SPECTROSCOPY

In AAS and AFS an auxiliary *light source* is required to produce the radiation which is to be absorbed (and partly re-emitted as fluorescence) by the analyte in the atomizer. (See Section 18.3.1.1.1.)

Notes: The traditional term "light" here also includes other than visible radiation, although according to the International Commission on Illumination (C.I.E. Publication No. 17, E.1.1, 1970), this term should be restricted to visible radiation (see footnote in Section 18.1 of this Part and Section 16.4.5 of Part I).

The general term "source" includes not only lamps but also the use of an auxiliary flame into which a constant amount of analyte vapour is introduced to produce the (primary) radiation.

Light sources are conveniently distinguished as *(spectral) continuum sources* or *(spectral) line sources*. Examples of spectral continuum sources used in AAS or AFS are the *tungsten-*

filament lamp and the *high-pressure zenon lamp*. These lamps radiate light as a result of the high temperature of the filament or as a result of a gas discharge. They belong to the class of *thermal radiators* (see also the definition of thermal radiation in Section 18.5.1.1).

Usually a spectral line source is employed containing atoms of the same element as the analyte. The line spectrum may often be superimposed upon a background that may be both discrete and/or a continuum.

The International Commission on Illumination (C.I.E.Publication No. 17, E.1.1, 1970) calls a spectral line source a "spectroscopic lamp".

Planck's law describes the spectral radiance B_λ^b (see Table 18.5.1) of a so-called full radiator (or black body) as a function of wavelength, λ, and temperature, T. A black body is a thermal radiator having an absorption factor $\alpha(\lambda)$ (see Table 18.5.1) equal to unity at all wavelengths. In thermal equilibrium, according to Kirchhoff's law, the spectral radiance B_λ of any radiator with $\alpha(\lambda) < 1$ is given by

$$B_\lambda = \alpha(\lambda) B_\lambda^b$$

where B_λ^b is to be taken at the temperature of the radiator.

Spectral line sources can be realized by an electrical discharge through a gas, a metal vapour, or a mixture of both at a low total pressure (the so-called *low-pressure discharge lamps*). The following types have found application in AAS and AFS as light sources.

In the *hollow-cathode lamp* accelerated electrons generate positive ions upon collision with atoms of the *carrier gas* (usually a noble gas). These ions gain energy in the electric field and collide with the cathode which usually has a hollow cylindrical form. Atoms of the cathode material are released by these collisions (*cathodic sputtering*). These atoms are excited in the discharge and radiate their spectral lines. These lamps can be made either with the cathode sealed within the same enclosure (*sealed lamp*) or the cathode element can be changed by dismantling the lamp (*demountable lamp*). The cathode material may be composed of only one element (*single-element lamp*) or of several elements (*multi-element lamp*).

Note: Carrier gas has also been called *fill-gas*, but this term is usually limited to sealed lamps.

Another common type of spectral line source is the *metal-vapour lamp* operated at low vapour pressures. The lamp is filled with a noble gas while the metal vapour is produced from the volatile element by the thermal effect of the discharge.

The *(high-frequency excited) electrodeless-discharge lamp* contains a noble gas at a low pressure and some volatile metal or metal salt (such as an iodide or chloride). A discharge produced in the noble gas by high-frequency fields generates electrons which by collisions excite the analyte atoms.

18.3.3 OPTICAL SYSTEMS

The functions of the optical system are to transfer, to select (spectrally, spatially, and temporally), and, possibly, to encode (by modulation) the radiation flux to be received by the photodetector. In addition, in AAS and AFS, the optical system should provide for the efficient conduction of the light beam through the analyte vapour contained in the atomizer (e.g. flame). In this Section, optical components, special systems, and their properties which are of interest in analytical flame spectroscopy will be considered. In Section 16.5 of Part I of this Nomenclature, general terms applicable to optical systems and optical instruments were considered. Those definitions will not be repeated here, but some of these terms are included in Table 18.3.2 for convenience, together with the recommended symbols.

The general term *spectrometer*, as it is used in this document (see Table 18.1), implies that quantitative measurements of intensities at one or more wavelengths are performed with a photoelectric detector. Spectral isolation of the desired radiation may be performed by means of *(optical) filters* (absorption filter, interference filter, etc.) or by a dispersing system. When the spectrometer is able to isolate only one narrow wavelength region, the instrument is called a *monochromator*. A *polychromator* is a multichannel spectrometer that allows spectral isolation of a number of narrow wavelength regions.

A *resonance spectrometer* consists of a "reservoir" of analyte atoms, which is irradiated by an external radiation beam containing, among others, the resonance frequencies of the analyte atoms. The vapour in the reservoir is excited specifically by this resonance radiation. The resulting fluorescence radiation, which is related to the intensity of the analyte resonance radiation in the original radiation beam, is measured.

In AAS, the sensitivity (see Section 18.4.2) may be improved by means of a *multipass system*, which allows the radiation beam from the light source to pass several times through the analyte vapour before it reaches the detector.

Most optical systems in use in flame spectroscopy are *single-beam systems*. For *double-beam systems* used in AAS, the radiation from the light source is split into the *sample beam* and *reference beam*.

The optical signal can be modulated or pulsed by the periodic interruption or intensity variation of the light beam of interest. *Light modulation* when combined with an a.c. measuring system provides some advantages. For example, in FES, the electrical signal arising from the flame emission may thus be distinguished from the d.c. dark current (see Section 18.3.4) of the photodetector. In AAS and AFS, modulation of the primary radiation beam (prior to its entrance into the flame), provides for the discrimination between the absorption or fluorescence signal and the thermal emission of the flame. Light modulation can be achieved by means of a mechanical *chopper*, or by modulating the electrical current through the lamp.

Table 18.3.2 contains some further terms and their symbols for measurable quantities connected with the optical system.

18.3.4 PHOTODETECTORS

The most commonly applied *photodetectors* are the *vacuum phototube* (without internal amplification), the *photomultiplier tube* (with internal amplification by secondary emission of electrons) and the *photovoltaic cell* (a semiconductor device producing an electromotive force upon irradiation).

Note: The photovoltaic cell is often called a barrier-layer cell or photodiode.

There are two components in the output current of a photodetector. One component, the *photocurrent*, i_f, is that portion induced by the radiation. The detector is said to be *linear* if the photocurrent is proportional to the radiant flux. The other component is called the *dark current*, i_d, because it continues to exist when the radiation flux is blocked.

The *responsivity* of a photodector (or in more general terms, the sensitivity) is the change in output current per unit change in radiant flux. The responsivity is wavelength dependent; the curve describing this dependence is called the *spectral response curve*.

Table 18.3.2 lists the quantitative terms and symbols related to photodetectors.

TABLE 18.3.2 Selection, detection, and readout of the analytical signal. Terms, symbols, and units for measurable quantities

Terms	Symbol	Practical unit	Note
Entrance-slit width of monochromator	s	mm	See Part I of this nomenclature series
Entrance-slit height of monochromator	h	mm	See Part I of this nomenclature series
Spectral bandwidth of monochromator	$\Delta\lambda_m$	nm	See Part I, Sections 16.2.8 and 16.5.2.2 respectively for the use of Å and choice of symbol
10%- or 1%-width of optical filter (measured between points of 10% or 1% of maximum transmission factor)	$\Delta\lambda_{0.1}$ or $\Delta\lambda_{0.01}$		See Part I, Sections 16.2.8 and 16.5.2.2 respectively for the use of Å and choice of symbol
Wavelength for maximum transmission of optical filter or monochromator	λ_m	nm	See Part I, Sections 16.2.8 and 16.5.2.2 respectively for the use of Å and choice of symbol
Optical conductance	G	cm² sr	For definition, see Part I, Section 16.5.3.2
Half-intensity width of source line	$\delta\lambda_s$	nm	See Part I, Sections 16.2.8 and 16.5.2.2 respectively, for the use of Å and choice of symbol. See Table 18.5.1 of Part III for definition of half-intensity width
Response time of a system (time needed to reach the reading that is a specified fraction, e.g. 99%, of the final value)	$\tau_r, \tau_{0.99}$	s	
Observation height (above top of burner)	h_{obs}	cm	
Frequency of light modulation	f_{mod}	Hz	
Electric current for operating light source	i_s	A	The official symbol I for current may give confusion with I for intensity
Dark current (of photodetector)	i_d	A	For definition see Section 18.3.4. The official symbol I for current may give confusion with I for intensity

TABLE 18.3.2 (*Continued*)

Terms	Symbol	Practical unit	Note
Photocurrent	i_f	A	For definition see Section 18.3.4. The official symbol I for current may give confusion with I for intensity
Solid angle over which emission is measured	Ω_E	sr	
Solid angle over which fluorescence is measured	Ω_F	sr	
Solid angle over which radiation is absorbed by flame from light source	Ω_A		

18.3.5 THE ELECTRICAL MEASURING SYSTEM

In the electrical measuring system the electrical system is delivered by the photodetector (see Section 18.3.4) and processed and converted to a *reading* on an appropriate readout device. The reading provides a measure (see Section 18.4.1) of the radiant flux emitted, absorbed, or fluoresced in the flame. In the absence of analyte, the signal trace recorded as a function of time or of wavelength is called the *baseline*. In order to speed up the evaluation of an analysis, the readout is sometimes directly calibrated in terms of concentration or amount of analyte. A non-linear analytical curve (see Section 18.4.2) may be linearized by means of a *curve corrector*. The background (see Section 18.4.1) can be compensated electronically by means of a *background corrector*.

Readings of small differences in two relatively large signals may often be made more readily by *scale expansion* of a segment of the reading range. When *zero suppression* is used, one of the signals is suppressed by displacing the zero of the meter, usually by electronic means, and the difference signal is increased by increasing the amplifier gain.

The *response time* (see also Table 18.3.2) is the time needed for the readout to reach a specified fraction (which must be stated) of its final value if the photodetector is suddenly exposed to the radiation flux. If this fraction is chosen to be $(1 - 1/e) = 0.63$, then the response time is called the *time constant*.

Scatter and drift will be discussed in Section 18.4.3.1.

18.3.6 SURVEY OF TERMS, SYMBOLS, AND UNITS FOR MEASURABLE QUANTITIES IN THE OPTICAL AND MEASURING SYSTEMS. (See Table 18.3.2)

18.4 TERMS AND SYMBOLS RELATING TO THE ANALYTICAL PROCEDURE AND THE PERFORMANCE OF AN ANALYSIS

18.4.1 GENERAL ANALYTICAL TERMINOLOGY IN FLAME SPECTROSCOPY

In analytical flame spectroscopy the *sample* may be a *solution* or is brought into solution. We distinguish between the *solvent* (e.g. an alcohol-water mixture), the *analyte*, i.e. the element sought, and the *concomitants*, i.e. any species other than the analyte and the solvent. Some concomitants may be present in known and constant concentrations through having been added during chemical pretreatment of the sample, and are called *additives*. Others, which were in the original sample, may have variable and/or unknown concentrations.

The border line between the solvent and the concomitants cannot always be sharply drawn. For example, when alcohol is added to the solution in constant known proportions (in order to improve nebulization or to promote chemiluminescence), it is regarded as a part of the solvent; when present in the original sample, it may be regarded as a concomitant.

A *reference solution* is a solution with the same solvent as the sample, and contains the analyte, and possibly some concomitants, in known concentrations.

A *blank solution* is a solution that does not intentionally contain the analyte, but in other respects has, as far as possible, the same composition as the sample solution. A *solvent blank* consists only of the solvent.

The *analytical result* is the final value of the *concentration*, c, or *quantity*, q, of the element sought, after all subprocedures and evaluations have been performed. For example, the analytical result may be obtained from a meter reading which provides a *measure* of some physical quantity, such as an analyte emission intensity, absorbance, or fluorescence intensity measured at a wavelength of an analytical line. The physical quantity carrying the information on the analyte concentration is called the *(analyte) signal*.

Note: If the position of the sensitivity knob of the instrument is changed, the reading, but not the measure, is changed.

The measure obtained when a blank solution is nebulized into the flame is called the *blank measure*, x_{bl}. The measure, x, obtained when the (reference) solution is nebulized, can be corrected for the blank measure, x_{bl}, either instrumentally or numerically. The difference, $x - x_{bl}$, is called the *net measure*. The signal that is observed when no solution or solvent is nebulized into the flame is called the *flame background* (*emission* or *absorption*). The background when a blank solution is nebulized in FES, FAS or FFS is called the *blank background* (*emission*, *absorption*, or *scattering*, respectively).

Note: See also Part I, Section 16.6, for a further explanation of some general terms.

18.4.2 ANALYTICAL CALIBRATION

The relationship between the measure, x, of the signal and the solution concentration, c, of the analyte is given by the *analytical curve*. This curve is generally established by making measurements on a series of reference solutions.

The derivative of the function $x = g(c)$, dx/dc, is the *sensitivity* of the analytical procedure. When the analytical curve is non-linear, the sensitivity is a function of the concentration.

In AAS, it is often desirable to compare slopes of analytical curves at low analyte concentration values for different lines and/or elements. This is often done by reporting concentration values of the analyte corresponding to 1% net absorption or 0.0044 absorbance. The term *characteristic concentration* is recommended for this particular concentration value.

Note: In the past the term 'sensitivity' has been incorrectly used for this concentration value.

Different techniques may be used for obtaining the analytical results. When the *analytical-curve technique* is used, the analytical result is read from an analytical curve covering the concentration range of interest.

When the *bracketing technique* is applied, the analytical result is found by graphical or numerical (usually linear) interpolation between measures of two reference solutions, one having a slightly lower, and the other a slightly higher analyte concentration than the unknown sample solution.

When the *(analyte) addition technique* is applied, successive known quantities of the analyte are added to aliquot portions of the sample solution. The net measures of the solutions thus obtained are plotted against the added concentrations. This plot is extrapolated to intercept the negative concentration axis. The analytical result is found from the corresponding concentration value.

Direct methods, i.e. those where the analyte produces the measured signal, are generally used in flame spectrophotometry. For some elements *indirect methods* have been applied. A given amount of another element, whose measured signal depends on the analyte concentration present, is added to the sample solution. If the change in the signal of the latter element is measured, a measure of the analyte concentration is obtained (e.g. the determination of phosphate by the addition and measurement of strontium). In some cases, the added element reacts with the analyte in the solution before nebulization, the resulting compound being separated. The remainder of the added element, or the amount of the added element that is separated because of compound formation, is then measured (e.g. the chloride content may be determined by the addition and measurement of silver).

Note: See also Part II for a further explanation of some general terms.

18.4.3 ASSESSMENT OF AN ANALYTICAL PROCEDURE

18.4.3.1 Measurement scatter.

The measure observed when a sample solution is analyzed is, in general, composed of two portions: the blank measure and the net measure (see Section 18.4.1). The net measure refers to the (analyte) signal (see Section 18.4.1) and is thus the useful, informative portion. The blank measure can be isolated and observed when a blank solution is treated in exactly the same way as the sample solution. Because the analyte should not be present in the blank solution, the blank measure contains no information about the analyte.

When replicate determinations are made on a given sample or blank, the measured values will not be constant, but will show *scatter*. If the causes of the scatter are not known (or considered), they cannot be controlled and therefore appear accidental. Blank scatter and signal scatter both contribute to the uncertainty of the analytical result.

The numerical value for the scatter may be taken as the *standard deviation*, which is determined from a sufficiently large number of repeated determinations. For a set of n measured values x_j the standard deviation, s, is, in general, defined and calculated by the formula:

$$s = \left[\sum_{j=1}^{n} (x_j - \bar{x})^2 / (n - 1) \right]^{1/2}$$

where

$$\bar{x} = \frac{1}{n} \sum_{j=1}^{n} x_j$$

is the average.

The scatter in general is a composite quantity produced by random fluctuations from different sources. If these random fluctuations in the measure x are independent, the formula for the propagation of *variance* (square of the standard deviation) can be applied to this case in the following symbolic form:

$$s_{bl}^2 = s_I^2 + s_{II}^2 + s_{III}^2 + \ldots$$

Typical sources of scatter in the blank and net measure are the following (not all independent):

Preparation of solution (accidental differences in weighing, adsorption, etc.):
Accidental external contamination of solution:
Variation of impurities in reagents:
Random fluctuations of nebulization and transport:
Random fluctuations of desolvation, atomization, and excitation:
Random fluctuations in the background:
Random fluctuations from the light source:
Uncontrolled variations in the optical system:
Random fluctuations in the photodetector, such as shot noise in the photocurrent, and low-frequency flicker noise arising from instabilities in the photocathode:
Electronic noise in the electrical measuring device, such as shot noise and thermal or Johnson noise which is due to the thermal agitation of charge carriers in resistors:
Drift in amplifiers:
Reading errors, accumulated rounding errors in calculations, calibration errors, etc.

In a complicated measuring system the contribution of these different sources can rarely be predicted theoretically. *Blank scatter*, s_{bl}, can only be determined by making blank analyses and then treating the measures, x_{bl}, statistically. It is therefore highly important that such a series of blank analyses should be planned carefully. All causes for random fluctuations involved in the original analytical procedure must play their full part, except those which are directly bound to the presence of the analyte.

18.4.3.2 Limit of detection, precision, and accuracy.

The merits of an analytical procedure may be characterized by its limits of detection, precision, and accuracy.

The *limit of detection* expressed as a concentration c_L, or quantity, q_L, is derived from the smallest measure, x_L, that can be accepted with confidence as genuine and is not suspected to be only an accidentally high value of the blank measure. The value of x_L is given by the equation:

$$x_L = \bar{x}_{bl} + k s_{bl}$$

where $\bar{x}_{bl}$ is the mean and s_{bl} the standard deviation of the blank measure, and k a numerical factor chosen according to the specified confidence level. The limit of detection is usually obtained directly from the analytical curve. A value of $k = 3$ is strongly recommended for the reasons given in Part II, Section 17.4.1. Although a value of 2 has often been used, this value is not recommended. To avoid ambiguity, the k-value should be indicated as follows: $x_{L(k=3)}$. The time constant of the measuring device should be stated specifically so that meaningful comparison between the limits of detection for different instruments may be made.

The *precision* of an analytical procedure can be conveniently expressed by the standard deviation, s, or by the relative standard deviation, $s_r = s/\bar{c}$ in the analytical result (see Part II, Section 17.2.4).

The *accuracy* relates to the difference between the analytical result obtained by a given analytical procedure and the (known or assumed) true analyte concentration in the sample.

The *bias* characterizes the systematic error in a given analytical procedure and is the (positive or negative) deviation of the mean analytical result from the (known or assumed) true value.

If materials with certified analyte concentration values (i.e. the best estimates of the true values) are not available, and the results cannot be compared with those obtained by other reliable methods, several test experiments may be used to check the accuracy of a procedure. The *recovery test* is based on the addition of a known amount of the analyte to the sample at an early stage of the analytical procedure. This amount must be found when the analytical result for the sample solution without addition is subtracted from the result for the sample solution with addition. In the *dilution test*, different known dilutions of the sample are made and the results are compared.

The correct outcome of these tests is a necessary, but not sufficient proof of the absence of systematic errors in a given analytical procedure.

Note: For a further explanation, see Part II and *Pure Appl. Chem.*, Vol. 18, No. 3 (1969), p. 439.

18.4.4 INTERFERENCES BY CONCOMITANTS

18.4.4.1 General.

The presence of concomitants in a sample can cause *interferences*, i.e. systematic errors in the measure of the signal. An interference may be due to a particular concomitant or to the combined effect of several concomitants. A concomitant causing an interference is called an *interferent*. When C is an interferent and A the analyte, there is interference of C on A. When an element X acts as an interferent on the analysis element Y, and Y acts as an interferent on the analysis element X, a *mutual interference* is said to exist.

Note: The influence of the solvent or of the flame background (without nebulization) on the measure is not called an interference. This is logical, because the sample and the reference solutions presumably contain the same solvent (see Section 18.4.1).

An interference will cause an error in the analytical result only if the interference is not adequately accounted for in the evaluation prodecure. The analyte concentration determined when the interference is not accounted for is called the *apparent concentration*. If the apparent concentration is larger than the true concentration, *enhancement* occurs or, if smaller, a *depression*. The *interference curve* relates the measure, or apparent concentration of the analyte, to the concentration of interferent, at fixed analyte concentration.

18.4.4.2 Classification of interferences.

18.4.4.2.1 Spectral interferences.

Spectral interferences are due to the incomplete isolation of the radiation emitted or absorbed by the analyte from other radiation detected by the instrument. Their occurrence may be established by comparing the measures of the analyte-free blank solution and the solvent blank (1).

Spectral interferences are usually strongly dependent on the spectral bandwidth of the monochromator. Spectral interference may arise:

In FES — from radiation (spectral continuum, molecular bands, or atomic lines, called *interfering lines*) emitted by the concomitants. Spectral interference may also arise from stray or scattered light or spectral ghosts that reach the detector:

— by the indirect effect of the concomitants on flame background (which is sometimes difficult to distinguish from their direct contribution to the background).

In FAS and/or FFS — by absorption or fluorescence of radiation by overlapping molecular or atomic lines of concomitants:

— by thermal emission of concomitants transmitted by the monochromator or received by the photodetector as stray light, when the light source is not modulated:

— by scattering of source radiation by nonvolatilized particles formed by the concomitants:

— by the indirect effect of the concomitants on the blank background absorption or scattering in the flame:

— by foreign line absorption and/or fluorescence if the corresponding radiation happens to be emitted by the light source, in addition to the analysis line, within the spectral bandwidth of the monochromator, particularly when a continuum source is used.

Note: The spectral interference found when a blank solution is nebulized need not be identical with the true spectral interference, since the presence of analyte in the sample might, in turn, influence of emission or absorption of the concomitants through one of the effects to be discussed in this section.

18.4.4.2.2 Non-spectral interferences.

For interferences other than spectral, the analyte signal itself is diectly affected. The *non-spectral interferences* may be classified according to the following viewpoints:

(a) To the place or stage at which the particular interference occurs, i.e. transport, solute-volatilization, vapour-phase and spatial-distribution interferences:

(b) To the effects on different elements, i.e. *specific* and *non-specific* interferences:

(c) To the properties which are decisive for the mechanism of the interference, i.e. *physical* and *chemical* interferences.

Note: The latter classification is discouraged since it can easily lead to confusion. Some physical processes (e.g. volatilization) determined by the physical properties of the particles formed by the analyte in the presence of the interferent depend upon the chemical properties of the analyte and interferent.

These different classifications do not exclude one another. If the interference cannot be specified, the term *effect* may be used. Thus the *matrix effect* is a composite interference due to all the concomitants, except for the additives (see Part I, Section 16.6.1); the *anion, cation,* or *organic effect* includes all interferences caused by the presence of different anions, cations, or organic constituents of the sample. When a particular solvent other than water is used, its effect on the signal (as compared with a aqueous solution) is not to be considered as an interference (see *Note* in Section 18.4.4.1).

Transport interferences affect the amount of desolvated sample passing through the horizontal flame cross-section per unit time at the observation height. They include factors affecting the rate of liquid consumption, F_1, the efficiency of nebulization, ε_n, and the fraction desolvated, β_s. They may be classified as non-specific (and physical).

Solute-volatilization interferences are due to changes in the volatilization rate of the dry aerosol particles in the case when volatilization of the analyte is incomplete in the presence and/or absence of the concomitant. These interferences can either be specific, if the analyte and interferent form a new phase of different thermostability, as when Mg and Al form $MgAl_2O_4$ in an air-acetylene flame, or non-specific, if the analyte is simply dispersed in a large excess of the interferent, as when Ag is dispersed in ThO_2. If the interferent has a high boiling point, this latter is sometimes referred to as a blocking interference. It is often difficult to make sharp distinctions between the specific and non-specific solute-volatilization interferences.

Note: Solute-volatilization interferences do not necessarily depress the signal. Effects due to compounds causing explosive disintegration of the solid aerosol particles and consequent enhancement also belong to this group.

Vapour-phase interferences are caused by a change in the fraction of analyte dissociated,[†] ionized, or excited in the gaseous phase. These interferences may be called *dissociation, ionization,* and *excitation interferences* respectively. An excitation interference may occur when the concomitant alters the flame temperature. Experimentally these interferences may be easily recognized because they take place even when twin nebulizers are used for aspirating the analyte and interferent separately. All interferences of this type are specific.

Spatial-distribution interference may occur when changes in concentration of concomitants affect the mass flow rates or mass flow patterns of the analyte species in the flame. If they are caused by changes in the volume and rise velocity of the gases formed by combustion, in extreme cases manifesting themselves by changes in the size and/or shape of the flame, they are non-specific and are called *flame-geometry interferences*. However, if caused by changes in diffusion processes they may be specific. Thus the *lateral diffusion interferences* arise when the presence of concomitants delays the vaporization of spray droplets or solid particles thereby shortening the time available for lateral diffusion of the analyte gaseous species before they reach the viewing field of the spectrometer.

18.4.4.3 Reduction of errors due to interference for given instrumental conditions.

Several techniques may be used to reduce or eliminate analytical errors resulting from various types of interferences. Apart from changing the instrumental conditions, the

---

[†] Here "dissociation" means the formation of free neutral atoms from free molecules in the gaseous phase (see also Section 18.6.1). The term atomization is here not appropriate because the latter also covers the formation of free atomic ions (see Section 18.3.1.1.1).

following techniques represent some of those in current use.

In the *reference-element technique*, the measure of the analyte is compared with the measure of a *reference element* (see Part I, Section 16.6.2.2). This technique is used mainly for minimizing non-specific interferences.

In the *analyte addition technique* (see Section 18.4.2) errors arising from both specific and non-specific interferences, but not from spectral interferences, are minimized.

In the *simulation technique*, reference solution sufficiently similar in quantitative composition to the sample solutions to be analyzed are used so that the interferences in the reference and sample solution are equivalent.

In the *buffer-addition technique*, an additive (called a *spectrochemical buffer*) is added to both the sample and reference solutions for the purpose of making the measure of the analyte less sensitive to variations in interferent concentration. Additives that may serve as spectrochemical buffers are:

Suppressors reduce emission, absorption, or light scattering by an interferent, thus removing or lowering spectral interference.

Releasers reduce solute-volatilization interferences by forming a compound preferentially with the interferent, thus preventing the analyte from entering a thermally stable compound.

Protective agents combine chemically with the analyte or interferent in such a way as to reduce the type of interferences discussed under Section 18.4.4.2.2.

Ionization buffers are added to increase the free-electron concentration by the flame gases, thus repressing and stabilizing the degree of ionization (see Section 18.6.2).

Volatilizers increase the fraction volatilized (see Section 18.3.1.2) either by forming more volatile compounds or by increasing the total surface area of all analyte particles (e.g. by explosive disintegration or by dispersal of the analyte in a highly volatile matrix).

Saturators are interferents added in sufficiently high concentration to the sample solution to reach the *saturation (plateau)* of the interference curve (see Section 18.4.4.1).

18.5 TERMS, SYMBOLS, AND UNITS RELATING TO RADIANT ENERGY, AND ITS INTERACTION WITH MATTER

Relevant descriptive terms relating to the emission, absorption, and fluorescence of optical radiation are discussed in this Section. Terms for measurable quantities are listed separately in a table, together with their recommended symbols and practical units.

18.5.1 DESCRIPTIVE TERMS RELATING TO THE EMISSION, ABSORPTION, AND FLUORESCENCE OF RADIATION

18.5.1.1 Emission.

The emission spectrum of a light source may consist of separate *spectral lines* (see Section 16.7.6.1 in Part I) and/or a *spectral continuum* with a continuous distribution of frequencies or wavelengths. Spectral lines emitted by free neutral atoms and ions are called *atomic lines* and *ionic lines* respectively (see Section 16.7.6.1 of Part I). The spectral lines emitted by free molecules are grouped together in *spectral bands*. When the resolution of the monochromator (see Section 16.5.5.5 in Part I) is insufficient, some bands may appear as a (quasi-) continuum in the recorded spectrum.

The emission of an atomic line is the result of a transition of an atom from a state of higher excitation to a state of lower excitation. When the lower state of the excitation is the ground state, the line is called a *resonance line*.

Note: Some handbooks define a resonance line as the line that originates from the lowest excitation state for which an optical transition to the ground state is allowed. Other handbooks adopt the wider definition presented in the text (compare, e.g. *Lexicon der Physik*, edited by H. Franke, Stuttgart, with the *Handbook of the American Institute of Physics*). The broader definition corresponds to common usage in analytical flame spectroscopy. If desired, the resonance line (or doublet) originating from the lowest excited level(s) may be specifically called the first resonance line (or doublet).

The radiation originating from a source where all particles are in a state of thermal equilibrium (see Section 18.6.1) is called *thermal radiation*. This term applies for the radiation of a spectral continuum as well as of isolated spectral lines or bands.

When the excited state from which the transition originates is mainly populated as a direct result of a chemical reaction, the radiation process is called *chemiluminescence*.

Note: The adjective "thermal" as such does not indicate the kind of process (collisional, chemical or radiative) that is responsible for the excitation of the radiating substance. Note, however, that the term chemiluminescence specifies the kind of excitation process. These two concepts therefore do not necessarily exclude each other. In flames, "thermal chemiluminescence" can exist, if the chemical species involved in the chemiluminescent reaction are in chemical equilibrium. *Suprathermal chemiluminescence* results if the concentrations of the chemical species taking part in the excitation reaction are above the equilibrium value.

18.5.1.2   Absorption and self-absorption.

When a light beam traverses a flame or other hot gases into which a sample is nebulized, its intensity (in the beam direction) may be attenuated by several processes (1). First, radiation may be lost due to (real) *absorption* (2). Furthermore, *scattering* by particles in the condensed (1) or gaseous phase (3) may change the direction (but not the energy, or frequency) of the incident photons. Resonance fluorescence (see Section 18.5.1.3) is a special case of scattering by free atoms or molecules. Finally, photons may be removed from the original beam when they are transformed into photons of different frequency and direction as a result of non-resonance fluorescence (see Section 19.5.1.3).

Note(1). In atomizers (see Section 18.3.1.1) which incorporate optical windows, trivial reflection losses may also occur. Random reflections of the light beam by unevaporated droplets in the flame are for practical reasons not distinguished here from scattering effects.

Note(2). Conversion of radiation into heat occurs when an atom that has been excited by photon absorption subsequently loses its excitation energy through collisional processes. The energy lost is converted to kinetic energy (see also Section 18.5.1.3). Note that the expression: "absorption of a radiation beam in a medium" means the indirect conversion of radiant energy into heat, whereas the expression "absorption of a photon by an atom" only means the transition of an atom to a higher excited level. When the excited atom returns to the lower level by re-emission of a photon of the same energy, no conversion of photon energy into heat results.

(3): Scattering by free atomic or molecular species is not important in flames when the frequency of the photon does not correspond to any of the optically allowed transitions in the atom or molecule.

The absorption of a photon, by which the atom is raised from a lower level to a higher one, is the reverse process of photon emission. Each atomic line appearing as an emission line can thus in principle also occur as an absorption line. However, since the overwhelming majority of the atoms in a flame are normally in the ground state, absorption of photons (whether followed by conversion into heat or by fluorescence) is usually detectable only with resonance lines (see Section 18.5.1.1) or with lines absorbed by atoms or ions in low-lying excited states.

For a similar reason, *self-absorption* (see Part I, Section 16.7.6.2) is usually found with the resonance lines emitted by the flame or spectral source. There is a certain probability that the photons of the resonance line which are generated inside the light source, e.g. a flame or a hollow-cathode lamp, may be absorbed on their way out by ground-state atoms and of being partly converted into heat. The loss will be larger, the thicker the cloud of atoms and/or the higher their concentration (4).

Self-reversal of emission lines is a special case of self-absorption which occurs when the radiating core inside the light source (flame, hollow-cathode lamp) is surrounded by a mantle of atomic vapour in which little or no excitation takes place (see also Part I, Section 16.7.6.3). Within a flame, this situation occurs when the temperature in the mantle is appreciably lower than at the centre. The absorption, in the mantle, of the line radiation from the core is then no longer (fully) compensated by the emission of the mantle itself. Since the absorption factor (see Table 18.5.1) has a peak value at the centre of the line, the uncompensated loss of radiation will here be more pronounced than in the line wings. This may result in the appearance of a minimum or *reversal dip* in the centre of the line profile. In the extreme case, when practically only the line wings remain, the lines may appear as two diffuse lines.

(4): At high concentrations, where self-absorption becomes noticeable, the intensity of a resonance emission line will therefore increase less proportionately with increasing atom concentration in the flame. If the emitting species occurs in a zone of the flame with a homogeneous radial temperature distribution, the relationship between line intensity and atom concentration is described by the *curve-of-growth*. This curve has a linear branch in the range of low concentrations where self-absorption is still negligible. At high concentrations, the intensity increases as the square-root of the concentration (square-root branch).

18.5.1.3 Fluorescence.

The absorption of photons from a primary beam to raise an atom to a higher excited state may be followed directly or indirectly) by (secondary) photon emission; this process is called *atomic fluorescence*. When the wavelengths of the absorbed radiation in the exciting beam and of the re-emitted radiation are identical *resonance fluorescence* is said to occur. Resonance fluorescence may be considered to be a special case of scattering (see Section 18.5.1.2). When the wavelengths of the two radiations are different, several cases can be distinguished. *Direct line fluorescence* exists when the transitions in the absorption and fluorescence process have a common upper level. When the upper levels are different, *stepwise line fluorescence* occurs. *Stokes* and *anti-Stokes fluorescence* apply when the wavelength of the fluorescence radiation is longer or shorter, respectively, than that of the absorbed radiation.

Note: In resonance fluorescence the line need not be a resonance line (see Section 18.5.1.1), although this will most often be the case.

In stepwise line fluorescence, atoms excited to the upper level by the primary beam are transferred, usually by collisions, to another excited level from which the fluorescent line is emitted.

TABLE 18.5.1 Radiant energy and its interaction with matter.
Terms, symbols, and units for measurable quantities

Terms	Symbol	Practical unit	Note
(Radiant) energy	Q	J	See also Part I, Section 16.4.
Spectral (radiant) energy	Q_λ	J nm^{-1}	When a quantity is considered per unit of wavelength (or frequency) at a given wavelength (or frequency), it is preceded by the adjective 'spectral' (See Part 1, 16.2.2). The appropriate symbol is obtained by adding λ (or ν) as a suffix, e.g. Q_λ. Note that the units for Q(J) and Q_λ (J nm^{-1}) are different. The radiant energy contained in a small wavelength interval, dλ, is given by Q_λ dλ.
Intensity (of radiation)	I	1	The general term "intensity" may be used as a loose relative expression referring to any radiant quantity without specification (see also Part I, Section 16.6.3.1). Note, however, that "radiant intensity" has a well-defined meaning.
Radiant intensity		W sr^{-1}	All of the above notes apply to this term.
Radiant flux	Φ	W	See also Part I, 16.4. When this quantity is considered per unit of wavelength (or frequency) at a given wavelength (or frequency), it is preceded by the adjective "spectral" (see Part I, 16.4.2). The appropriate symbol is obtained by adding λ (or ν) as a suffix, e.g. Φ_λ. Note that the units for Φ(W) and Φ_λ (W nm^{-1}) are different. The radiant flux contained in a small wavelength interval, dλ, is given by Φ_λ dλ.
Radiant flux incident on (absorbing) medium	Φ_o	W	See also Part I, Section 16.5.
Radiant flux transmitted by (absorbing) medium	Φ_t	W	See also Part I, Section 16.5.
Radiant flux absorbed by medium	Φ_a	W	See also Part I, Section 16.5.

TABLE 18.5.1 (*Continued*)

Terms	Symbols	Practical unit	Note
Transmission factor (Φ_t/Φ_o)	τ	1	See also Part I, Section 16.5. This quantity also includes the transmission properties of the apparatus (cell windows, etc.). The value of this quantity, for strictly monochromatic radiation of wavelength λ, is denoted by $\tau(\lambda)$ (not: τ_λ).
Internal transmission factor	τ_i	1	The value of this quantity, for strictly monochromatic radiation with wavelength λ, is denoted by $\tau_i(\lambda)$. This quantity refers to the transmission properties of the sample alone.
Absorption factor (Φ_a/Φ_o)	α	1	See also Part I, Section 16.5. The value of this quantity, for strictly monochromatic radiation with wavelength λ, is denoted by $\alpha(\lambda)$ (not: α_λ). When the absorption properties of the apparatus, e.g. cell windows, are excluded, the adjective "internal" is added. The term "absorptance" is not recommended in this context because of confusion with "absorbance".
(Internal) absorbance $(-\log_{10}\tau_i)$	A	1	This quantity refers to the transmission properties of the sample alone. In AAS however, the adjective "internal" is usually dropped.
Peak value of A (at the absorption line centre λ_o)	$A(\lambda_o)$	1	
Absorption path length	$l \ldots b$	cm	
Integral absorption (of spectral line) $\int \alpha(\lambda)d\lambda$		nm	For the use of $\overset{o}{A}$, see Part I, Section 16.2.8. This quantity describes the energy absorbed from a continuum within the wavelength profile of a spectral line.
(Radiant) energy density	u, w	J cm^{-3}	See also Part I, Section 16.4. When this quantity is considered per unit of wavelength (or frequency) at a given wavelength (or frequency), it is preceded by the adjective "spectral" (see Part I, 16.4.2.). The appropriate symbol is obtained by adding λ (or ν) as a suffix, e.g. u_λ.
Radiance	$B \ldots L$	W sr^{-1} cm^{-2}	See also Part I, Section 16.4. When this quantity is considered per unit of wavelength (or frequency) at a given wavelength (or frequency), it is preceded by the adjective "spectral" (see Part I, 16.4.2). The appropriate symbol is obtained by adding λ (or ν) as a suffix, e.g. B_λ.
(Einstein) transition probability for spontaneous photon emission (by optical transition from upper state u to lower state l)	A, A_{ul}	s^{-1}	Also called the (Einstein) coefficient for spontaneous emission.
Oscillator strength for absorption (by optical transition from lower state l to upper state u) f-value	f, f_{lu}	1	
Intensity of spectral line due to transition from the upper state u to lower state l.	I_{ul}	1	

TABLE 18.5.1 (*Continued*)

Terms	Symbols	Practical unit	Note
Wavelength of (atomic) line centre	λ_o	nm	For the use of Å see Part I, Section 16.2.8.
Quantum efficiency of fluorescence (number of photons re-emitted per second/number of primary photons absorbed per second	Y, Y_q	1	
Power efficiency of fluorescence (radiant flux re-emitted/primary radiant flux absorbed	Y_p	1	
Total quantum efficiency of fluorescence (= for the case when the upper level of the fluorescence transition is populated directly or indirectly (by 2-step process) by absorption of more than 1 spectral line	Y_t	1	
Half-intensity width (full width at half peak height of a spectral line profile)	$\delta\lambda$	nm	The term half-width is sometimes used instead of half-intensity width, but may readily be misunderstood as half the full width. For the use of Å, see Part I, Section 16.2.8.
Doppler half-intensity width (of spectral line due to Doppler broadening)	$\delta\lambda_D$	nm	The term half-width is sometimes used instead of half-intensity width, but may readily be misunderstood as half the full width. For the use of Å, see Part I, Section 16.2.8.
Collisional half-intensity width (of spectral line due to collisional broadening)	$\delta\lambda_C$	nm	The term half-width is sometimes used instead of half-intensity width, but may readily be misunderstood as half the full width. For the use of Å, see Part I, Section 16.2.8.
α-parameter ($\sqrt{(\ln 2)}(\delta\lambda_C/\delta\lambda_D)$)	α	1	Also called: line-broadening parameter or damping constant. In this definition of α, the natural line-broadening is disregarded. When natural line-broadening is important, it should be included in the numerator.

Collisions of fluorescing atoms with other atoms or molecules are said to *quench* the fluorescence when they destroy the state of excitation brought about by absorption of the primary photons. The number of secondary photons will then be smaller than the number of primary photons absorbed. The extent of quenching is determined by the competition between the rates of radiative and collisional de-excitation of the excited atoms, and quantitatively expressed by the *efficiency of fluorescence* (for definition see Table 18.5.1).

18.5.2   TERMS, SYMBOLS, AND UNITS FOR MEASURABLE QUANTITIES

Table 18.5.1 presents terms with their symbols and practical units for some measurable quantities belonging to this Section. Section 16.4 of Part I lists additional terms. Although practical units generally conform, alternative symbols that differ from those in Part I are occasionally recommended.

18.6 TERMS, SYMBOLS, AND UNITS RELATING TO THE GASEOUS STATE OF MATTER

Analytical flame spectroscopy and similar flame techniques are based on the interaction of radiation with the analyte from the sample. The strength of this interaction depends on the properties and state of the analyte in the vapour phase. In the following, we restrict ourselves to this phase (the transformation of the analyte from the condensed phase into the vapour phase has already been discussed in Section 18.3.1). A few descriptive terms will be mentioned, followed by a list of terms for measurable quantities with their symbols and practical units.

18.6.1 DESCRIPTIVE TERMS CONCERNING THE GASEOUS STATE OF MATTER

In *(thermodynamic) equilibrium* the state of a system is generally described by the *thermodynamic* or *absolute temperature* (Table 18.2.1) which occurs as a universal parameter in the distribution laws. These laws determine the state of *excitation, ionization,* and *dissociation* in which the analyte is found in the gaseous phase.

Note: The Maxwell-Boltzmann law describes the distribution over the translational and internal energies of the particles. Planck's law describes the distribution of the radiant energy over the spectrum. Saha's law describes the distribution of the particles over their different states of ionization, while the mass action law determines the fraction of element that is bound in molecular form.

In atomic spectroscopy the term "dissociation" usually refers to the splitting of the free molecule into fragments, one or more of which is a free neutral analyte atom, and is thus important in connection with atomization (see Section 18.3.1.1.1).

Strictly speaking, thermodynamic equilibrium implies that the spectral energy density obeys Planck's law (see Note in Section 18.3.2). In flames of analytical interest, this condition is not met except at the centre of strongly self-absorbed resonance lines. However, radiative processes usually play a minor part in the population of the excited states, at least if the flame is not irradiated by an external light source. The distribution of the particles over their various energy states and over the various forms in which they may occur (as neutral or ionized atoms, or as molecules) is then hardly affected by deviation from Planck's law. The system is then said to be in *thermal equilibrium*. This implies that there is *chemical equilibrium* between all chemical species (including ions and electrons) in the system, as well as *physical equilibrium* for the translational and internal energies of the particles. However, there need then be no *radiative equilibrium*.

Note: In the absence of radiative equilibrium, it is still meaningful to define thermal radiation as has been done in Section 18.5.1.1.

18.6.2 TERMS, SYMBOLS, AND UNITS FOR MEASURABLE QUANTITIES

Many of the terms listed in Table 18.6.1 are also found in Part I, Section 16.7.3, where the terms are discussed in more detail.

TABLE 18.6.2 Properties and state of matter.
Terms, symbols, and units for measurable quantities

Terms	Symbols	Practical units	Note
Excitation energy	E_{exc}	J...eV	See Part I, 16.7.4. The electron volt (eV), as a practical unit for energy on an atomic scale, is still in common use. When several excited states are to be distinguished, the excitation energy of state q may be written as E_q. When several kinds of particles are also considered, the excitation energy of state q of a particle X may be written as $E_q(X)$.
Excitation potential	V_{exc}	V	See Part I, Section 16.7.4.
Ionization energy	E_{ion}, E_i	J...eV	See Part I, Section 16.7.4.
Ionization potential	V_{ion}, V_i	V	See Part I, Section 15.7.4.
Dissociation energy	E_{dis}, D_o, D_{XY}	J...eV	This is the minimum energy required to dissociate one molecule of XY at zero K in perfect gas state. The fragments must be specified when the molecule contains more than 2 atoms. The molar dissociation energy refers to one mole of substance XY. In chemistry, the symbol D_o is customarily used.
Dissociation potential	V_{dis}	V	
Degree of ionization $M^+/(M + M^+)$		1	M denotes an atom; M denotes number density of M in the flame (see also Table 18.2.1).
Degree of dissociation $M/(M + MX)$		1	M denotes an atom; M denotes number density of M in the flame (see also Table 18.2.1).
Statistical weight of particle in state q	g_q	1	When several kinds of particles are considered, the symbol may be extended by adding the chemical symbol as in: $g_q(X)$.
Statistical weight of particle in ground state	g_o	1	When several kinds of particles are considered, the symbol may be extended by adding the chemical symbol as in: $g_o(X)$.
Partition function	Z, Q	1	When several kinds of particles are considered, the symbol may be extended by adding the chemical symbol as in: $Z(X)$.
(Number) density of particles in state q (see also Table 18.2.1)	n_q	cm^{-3}	When several kinds of particles are considered, the symbol may be extended by adding the chemical symbol as in: $n_q(X)$.
(Number) density of particles in ground state	n_o	cm^{-3}	When several kinds of particles are considered, the symbol may be extended by adding the chemical symbol as in: $n_o(X)$.
(Number) density of free atoms	n_{at}, n_a, n_M $n(M), M$		
(Number) density of free ions (M^+)	n_{ion}, n_i, n_{M^+} $(M^+); M^+$	cm^{-3}	
(Number) density of free electrons	n_e, e^-	cm^{-3}	
(Number) density of molecules MX	n_{MX}, MX	cm^{-3}	
Total (number) density of element M $(n_M + n_{M^+} + n_{MX})$	$n_t, (n_t)_M$	cm^{-3}	

TABLE 18.6.2 (*Continued*)

Terms	Symbols	Practical units	Note
Ionization constant ($n_i n_e/n_a$ in equilibrium at T)	$K_i, K_i(T)$	cm^{-3}	
Dissociation constant ($n_M n_X/n_{MX}$ in equilibrium at T)	$K_d, K_d(T)$	cm^{-3}	

18.7 INDEX OF TERMS[†]

Absolute temperature (18.6.1 [‡]; Table 18.2.1)
absorption (18.5.1.2)
accuracy (18.4.3.2)
additive (18.4.1)
 aerosol (18.3.1.1.1)
analyte (18.4.1)
(analyte) addition technique (18.4.2, 18.4.4.3)
(analyte) signal (18.4.1)
analytical curve (18.4.1)
analytical curve technique (18.4.2)
analytical result (18.4.1)
angular nebulizer (18.3.1.1.2)
anion effect (18.4.4.2.2)
anti-Stokes fluorescence (18.5.1.3)
apparent concentration (18.4.4.1)
aspiration (18.3.1.1.1)
atomic fluorescence (18.5.1.3)
atomic line (18.5.1.3)
atomization (18.3.1.1.1)
atomizer (18.3.1.1.1)

Background corrector (18.3.5)
baseline technique (18.4.4.3)
bias (18.4.3.2)
blank background (18.4.1)
blank measure (18.4.1)
blank scatter (18.4.3.1)
blank solution (18.5.1)
bracketing technique (18.4.2)
buffer-addition technique (18.4.4.3)
Bunsen-burner (18.3.1.1.3)
burner (18.3.1.1.3)

Carbon atomizer (18.3.1.3.3.1)
carbon-cup atomizer (18.3.1.3.3.1)
carbon-tube furnace (18.3.1.3.3.1)
carrier gas (18.3.2)
cathodic sputtering (18.3.1.3.3.2 and 18.3.2)
cation effect (18.4.4.2.2)
chamber-type nebulizer (18.3.1.1.2)
characteristic concentration (18.4.2)
chemical equilibrium (18.6.1)
chemical interference (18.4.4.2.2)
chopper (18.3.3)
concentration (18.4.1)
concentric nebulizer (18.3.1.1.2)
concomitant (18.4.1)
controlled flow nebulizer (18.3.1.1.2)
cooled hollow cathode (18.3.1.3.3.2)
current-carrying plasma (18.3.1.3.2)
current-free plasma (18.3.1.3.2)
curve corrector (18.3.5)
curve-of-growth (18.5.1.2)

Dark current (18.3.4)
demountable lamp (18.3.2)
depression (18.4.4.1)
desolvation (18.3.1.1.1)
dilution test (18.4.3.2)
direct-injection burner (18.3.1.1.3)
direct line fluorescence (18.5.1.3)
direct methods (18.4.2)
dissociation (18.6.1)
dissociation interference (18.4.4.2.2)
double-beam system (18.3.3)

[†]Quantitative terms and some qualitative terms that are defined only in the tables are not included in this index. They are easily found in the tables under the appropriate heading.

[‡]The numbers refer to the relevant sections.

drop generator (18.3.1.1.2)
dry aerosol (18.3.1.1.1)

Effect (18.4.4.2.2)
efficiency of atomization (18.3.1.2)
efficiency of fluorescence (18.5.1.3; Table 18.5.1)
efficiency of nebulization (18.3.1.2)
electrical flame-like plasma (18.3.1.3.2)
electrical measuring system (18.3.5)
electrodeless-discharge lamp (18.3.2)
electrodeless plasma (18.3.1.3.2)
emission (18.5.1.1)
enhancement (18.4.4.1)
excitation (18.6.1)
excitation interference (18.4.4.2.2)

Fill-gas (18.3.2)
filter (18.3.3)
flame (18.1)
flame background (18.4.1)
flame-geometry interference (18.4.4.2.2)
flash-back (18.3.1.1.3)
fluorescence (18.5.1.3)
fraction atomized (18.3.1.2)
fraction desolvated (18.3.1.2)
fraction volatilized (18.3.1.2)
fuel (18.3.1.1.3)
fuel-rich flame (18.3.1.1.3)

Graphite-cup atomizer (18.3.1.3.3.1)
graphite-rod furnace (18.3.1.3.3.1)
graphite-tube furnace (18.3.1.3.3.1)
gravity-fed nebulizer (18.3.1.1.2)

High-pressure zenon lamp (18.3.2)
hollow-cathode discharge (18.3.1.3.3.2)
hollow-cathode lamp (18.3.2)
hot hollow cathode (18.3.1.3.3.2)

Indirect methods (18.4.2)
inductively-coupled plasma (18.3.1.3.2)
inner zone (18.3.1.1.3)
interconal zone (18.3.1.1.3)
interference curve (18.4.4.1)
interference (18.4.4.1)
interferent (18.4.4.1)
interfering line (18.4.4.2.1)
interzonal region (18.3.1.1.3)
ionic line (18.5.1.1)

ionization (18.6.1)
ionization buffer (18.4.4.3)
ionization interference (18.4.4.2.2)

Laminar (flame) (19.3.1.1.3)
laser (18.3.1.3.3.3)
lateral diffusion interference (18.4.4.2.2)
light modulation (18.3.3)
light source (18.3.2)
limit of detection (18.4.3.2)
linear (detector) (18.3.4)
local analysis (18.3.1.3.3.3)
long tube device (18.3.1.3.1)
low-pressure discharge lamp (18.3.2)

Matrix effect (18.4.4.2.2)
measure (18.4.1)
Méker-burner (18.3.1.1.3)
metal-vapour lamp (18.3.2)
mist (18.3.1.1.1)
monochromator (18.3.3)
multi-element lamp (18.3.2)
multipass system (18.3.3)
multislot burner (18.3.1.1.3)
mutual interference (18.4.4.1)

Nebulization (18.3.1.1.1)
nebulizer (18.3.1.1.2)
mebulizer with heated spray chamber (18.3.1.1.2)
net measure (18.4.1)
non-specific interference (18.4.4.2.2)
non-spectral interference (18.4.4.2.2)

Organic effect (18.4.4.2.2)
outer zone (18.3.1.1.3)
oxidant (18.3.1.1.3)

Photocurrent (18.3.4)
photodetector (18.3.4)
photomultiplier tube (18.3.4)
photovoltaic cell (18.3.4)
physical equilibrium (18.6.1)
physical interference (18.4.4.2.2)
plasma jet (18.3.1.3.2)
pneumatic nebulizer (18.3.1.1.1 and 3.1.1.2)
precision (18.4.3.2)
premix burner (18.3.1.1.3)
polychromator (18.3.1.1.3)
primary-combustion zone (18.3.1.1.3)

protective agent (18.4.4.3)
pulse-discharge lamp (18.3.1.3.3.3)

Quantity (18.4.1)
quenching (18.5.1.3)

Radiative equilibrium (18.6.1)
rate of liquid aspiration (18.3.1.2)
rate of liquid consumption (18.3.1.2)
reading (18.3.5)
recovery test (18.4.3.2)
reference beam (18.3.3)
reference element (18.4.4.3)
reference-element technique (18.4.4.3)
reference solution (18.5.1)
reflux nebulizer (18.3.1.1.2)
relative standard deviation (18.4.3.2)
releaser (18.4.4.3)
resistance-heated device (18.3.1.3.3.1)
resonance fluorescence (18.5.1.3)
resonance line (18.5.1.1)
resonance spectrometer (18.3.3)
response time (18.3.5)
responsivity (18.3.4)
reversal dip (18.5.1.2)
reversed direct-injection burner (18.3.1.1.3)

Sample (18.4.1)
sample beam (18.3.3)
sampling boat (18.3.1.3.1)
sampling cup (18.3.1.3.1)
sampling loop (18.3.1.3.1)
saturation (plateau) (18.4.4.3)
saturator (18.4.4.3)
scale expansion (18.3.5)
scatter (18.4.3.1)
scattering (18.5.1.2)
sealed lamp (18.3.2)
secondary-combustion zone (18.3.1.1.3)
self-absorption (18.5.1.2)
self-reversal (18.5.1.2)
sensitivity (18.4.2)
separated flame (18.3.1.1.3)
shielded flame (18.3.1.1.3)
simulation technique (18.4.4.3)
single-beam system (18.3.3)
single-electrode plasma (18.3.1.3.2)
single-element lamp (18.3.2)

slot burner (18.3.1.1.3)
solute-volatilization interference (18.4.4.2)
solvent blank (18.3.5)
spatial-distribution interference (18.4.4.2.2)
specific interference (18.4.4.2.2)
spectral band (18.5.1.1)
spectral continuum (18.5.1.1)
(spectral) continuous source (18.3.2)
spectral interference (18.4.4.2.1)
(spectral) line source (18.5.1.1)
spectral line (18.5.1.1)
spectral response curve (18.3.4)
spectrochemical buffer (18.4.4.3)
spectrometer (18.3.3; Table 18.1.1)
spectrometry (Table 18.1.1)
spray chamber (18.3.1.1.2)
sprayer (18.3.1.1.1)
standard deviation (18.4.3.1)
stepwise line fluorescence (18.5.1.3)
Stokes fluorescence (18.5.1.3)
suction nebulizer (18.3.1.1.2)
suppressor (18.4.4.3)
suprathermal chemiluminescence (18.5.1.1)

T-tubes (18.3.1.3.1)
thermal equilibrium (18.6.1)
thermal radiation (18.5.1.1)
thermodynamic equilibrium (18.6.1)
thermodynamic temperature (18.6.1;Table 18.2.1)
three-slot burner (18.3.1.1.3)
time constant (18.3.5)
transport interference (18.4.4.2.2)
tungsten-filament lamp (18.3.2)
tubulent (flame) (18.3.1.1.3)
twin nebulizer (18.3.1.1.2)

Ultrasonic nebulizer (18.3.1.1.2)

Vacuum phototube (18.3.4)
vapour-phase interference (18.4.4.2.2)
variance (18.4.3.1)
volatilization (18.3.1.1.1)
volatilizer (18.4.4.3)

Zero suppression (18.3.5)

19. CLASSIFICATION AND NOMENCLATURE OF ELECTROANALYTICAL TECHNIQUES*

19.1 INTRODUCTION

Almost without exception, the recommendations made here are descriptive rather than prescriptive, in the sense that they reflect what seem to be the best — the most accurate, informative, and logical — of the names that have gained some currency in the prior literature. Many of these techniques, probably including polarography itself, would be unlikely to be given the names here recommended if they were just being developed now. Nevertheless their prior histories provide ample proof that those names are much too firmly established, both in the literature and in the minds of their users, to be dislodged. An attempt to develop a completely systematic and consistent nomenclature *ab initio* therefore seems futile and has not been made.

This report follows the general lines of its predecessor, but deviates from it in a number of details. A slightly different classification is given in Table 19.1, and the other Tables are differently arranged accordingly. Many techniques that have been developed, or that have become important in analysis or fundamental research, since 1960 have been added, and a few older or less important techniques are also included for the sake of completeness.

As much of the nomenclature recommended by Delahay, Charlot, and Laitinen has been retained as seemed possible, but several deviations therefrom are to be noted. The term "polarographic titration" has not been accepted in place of either "amperometric titration" or "polarometric titration" and is accordingly withdrawn. The terms "biamperometric" and "bipotentiometric" are withdrawn, although they have been more widely used than "polarographic titration" because most of the colleagues we consulted thought them to be objectionable on the ground that they seemed to denote measurements of two currents or two potential differences, respectively. Their replacements, "amperometry with two indicator electrodes" and "potentiometry with two indicator electrodes", contain more syllables, but it is hardly possible to suppose that both conciseness and reasonable accuracy can be achieved for any but a few of the techniques covered here.

It is recommended that the term "polarography" be used to denote the study of relationships between electric current and applied e.m.f. or electrode potential with a liquid electrode whose surface is periodically or continuously renewed. The most common polarographic

---

* Based on the approved Recommendations published in *Pure and Applied Chemistry*, Vol. 45, No. 2 (1976), pp. 81 - 97 which revises and brings up to date recommendations made by P.Delahay, G.Charlot and H.Laitinen in *Analytical Chemistry*, Vo. 32, No. 6 (1960), p. 103A et seq.

indicator electrode is the classical dropping mercury electrode, but this definition also comprises the use of dropping electrodes of other metals or liquid conductors; of variants such as multiple dropping electrodes and fritted discs from which droplets of the liquid conductor emerge into the solution being investigated; and streaming metal (or other liquid conductor) electrodes. It excludes the use of all stationary and solid electrodes, such as hanging drops and pools, regardless of the material from which these are made; it is recommended that the term "voltammetry" be used to denote the study of relationships between electric current and applied e.m.f. or potential with indicator electrodes of these types.

TABLE 19.1 Classification of electroanalytical techniques

---

1. Techniques in which neither the electrical double layer nor any electrode reaction need be considered (Table 19.2).

2. Techniques that involve double-layer phenomena but in which any electrode reactions need not be considered (Table 19.3).

3. Techniques involving electrode reactions

   A. Techniques involving electrode reactions and employing constant excitation signals (Table 19.4).

   B. Techniques involving electrode reactions and employing variable excitation signals

   (i) Variable excitation signals of large amplitude (usually considerably larger than $2 \times 2.3RT/F$ volt, approximately 0.12 V at 25°C) (Table 19.5).

   (ii) Variable excitation signals of small amplitude (usually considerably smaller than $2.3RT/F$ volt, approximately 0.06 V at 25°C) (Table 19.6).

Electroanalytical techniques have had some problems of symbology in recent years. Symbols recommended for quantities peculiar to individual techniques — such as the diffusion current constant in polarography and the transition time in chronopotentiometry — are not included here; they will be the subject of a separate report. However, IUPAC has recently published the second edition of *Manual of Symbols and Terminology for Physicochemical Quantities and Units* (Butterworths, London: 1975) prepared by the Physical Chemistry Division. Some of the symbols therein prescribed — such as V for volume, c for the concentration of a solute, and G for electric conductance — are used here without comment. Others, which conflict with traditional usage, must be the subject of further discussion with the object of achieving as high a degree of conformity between different disciplines as possible. For example, the symbol I for current, recommended by the *Manual* and by other international organisations, has been added in parentheses after each i, which is the symbol far more generally used in the electroanalytical literature. In addition, the quantity E (potential difference) is often called simply "potential" by analytical chemists.

The Tables are generally self-explanatory; it need only be remarked that graphical representations of typical measured responses and of all variable excitation signals are given for ease of classification and comparison. In general, only those techniques are included that have achieved reasonable analytical significance, or that seemed to present special problems of nomenclature. Similar representations have been given by many authors, of whom to our knowledge C. N. Reilley was the first.

Some general recommendations and comments deserve special mention.

19.1.1 There is widespread, but unfortunately not quite universal, agreement that the term "differential" should be used to denote measurement of a difference while "derivative" should be used to denote measurement of a rate of change, and these meanings are generally employed here. Thus a "differential potentiometric titration" is a titration that involves monitoring the difference between the potentials of two indicator electrodes (in two different solutions), while a "derivative potentiometric titration" is a titration that involves measuring, recording, or computing the first derivative of the potential of a single indicator electrode with respect to the volume or otherwise added amount of reagent. As the term "differential" has occasionally been given the meaning reserved here for "derivative", some authors have been driven to use "subtractive" in its place. It is recommended that the term "subtractive" be dropped, that "differential" be used in its place and that derivative techniques be so designated.

19.1.2 Some techniques, including polarography, have variants in which three- electrode configurations (including the indicator or working electrode, the auxiliary or counter electrode, and a reference electrode) are employed and in which some instrumental compensation for the ohmic potential drop is applied. In principle these techniques are equivalent to the corresponding ones with two-electrode configurations when proper corrections are made to the values of the applied e.m.f. in the latter. We have therefore drawn no distinction between them in these tables, but would encourage the use of terms such as "controlled-potential polarography" when the difference of instrumentation or configuration is to be emphasized. In general, variants of other techniques should be similarly named. For example, the application of a constant current to a rotating disc electrode and the observation of the dependence of potential on time should not be called chronopotentiometry without qualification because, contrary to what is specified in Section 19.4.12 below, it does not involve the use of a stationary electrode in an unstirred solution. It should instead be called "rotating-disc-electrode chronopotentiometry," thereby conveying the natures of the excitation signal and the dependence observed and also the difference between its mathematical foundations and those of ordinary chronopotentiometry. For reasons of space, only a few illustrations (e.g. Sections 19.4.20 and 19.4.22) are given below.

19.1.3 Different techniques for reagent addition are not distinguished below, but terms like "potentiometric weight titration" are available when needed.

19.1.4 Among the family of techniques based on cathodic or anodic stripping of (usually) previously deposited compounds or elements, only electrography is individually distinguished below. Other techniques of this family are best designated by names such as " anodic stripping voltammetry (*or* anodic stripping chronoamperometry with linear potential sweep", "anodic stripping controlled-potential coulometry", and the like. The term "stripping analysis" is widely used in the literature to denote the first of these, but would better be reserved as a generic name for the entire family of electroanalytical techniques based on stripping procedures.

19.1.5 As applied to triangular-wave and related techniques, the ungainly term "multicyclic" is avoided by recommending the use of "triangular-wave polarography (or triangular-wave voltammetry)" to denote the examination of only a single cycle, and prefixing this with the word "cyclic" to denote iteration or reiteration of the cycle.

19.1.6   In most electroanalytical techniques there is one electrode that serves as a transducer, responding to the excitation signal and the composition of the solution being investigated but without effecting any appreciable change of bulk composition over the ordinary duration of a measurement. This electrode is the "indicator electrode" or "test electrode", only the former term being used below to save space. When, however, the technique depends on effecting significant changes of bulk composition by the flow of current through the cell, this electrode is called the "working electrode". It is immaterial whether the change in bulk composition occurs in the solution phase or in the liquid metal constituting the working electrode. For example, a stirred mercury-pool electrode used for voltammetry (Section 19.5.9) is considered to be an indicator electrode, but if used for controlled potential coulometry (Section 19.4.27) it is considered to be a working electrode regardless of whether the electroreducible- or oxidizable substance under study is initially dissolved in the mercury or in the solution.

19.1.7   Throughout these tables the term "alternating" (current or voltage) denotes the use of a sinusoidal waveform.

19.1.8   Several recently proposed techniques may be regarded as close descendants of others appearing below but differ from them in the nature of the data-handling involved (as, indeed, may be said of amperometry and coulometry). In choosing names for these and others of the same sort, we recommend deriving them from names given below. Two typical examples would be "semi-integral polarography" and "convolution-integral linear-sweep voltammetry".

19.1.9   Some techniques, like conductometry (19.2.1) and differential potentiometry (Section 19.4.2), employ two indicator electrodes, but more often the second electrode is more or less non-polarizable and is used merely to complete the measuring circuit and to provide a suitably constant potential. An electrode serving these purposes is called a "reference electrode". Sometimes these functions of the reference electrode are separated by using a three-electrode configuration. This comprises (i) and indicator (or test) or working electrode; (ii) a reference electrode, through which no significant current is allowed to flow and which serves to permit observation or control of the potential of the first electrode; and (iii) an "auxiliary electrode" or "counter electrode", which serves to carry the current that passes through the first electrode.

Table 19.7 is an index to Tables 19.2 to 19.6. For each of the techniques included in Tables 19.2 to 19.6 index Table 19.7 lists all the names under which it may be sought, including both the names that are recommended and others that are not. Each entry in Table 19.7 includes a reference in the form 19.x.y where the number 19.x gives the number of the table in which the technique appears and y is its ordinal number in that table. These key numbers also appear in the first columns of Tables 19.2 - 19.6 to facilitate both finding techniques sought through the index of Table 19.7 and occasional cross-references within tables.

TABLE 19.2 Techniques in which neither the electrical double layer nor any electrode reaction need be considered

Key no.	Excitation signal	Independent variable	Measured response	Recommended name of technique	Typical response curve	Remarks
19.2.1	Alternating voltage; frequency $f <$ ca. 0.1 MHz	Concentration c	Conductance G	Conductometry		Dc conductometry is rarely employed but should be so designated. The spelling "conductimetry" is not recommended.
19.2.2		Volume V (or otherwise measured amount) of a reagent	Conductance G	Conductometric titration		
19.2.3	Alternating voltage, frequency $f >$ ca. 0.1 MHz	Concentration c	Conductance G, susceptance B, or admittance Y	High-frequency conductometry		The recommended term is inexact when B or Y is measured, but names like "susceptometry" cannot be encouraged.
19.2.4		Volume V (or otherwise measured amount) of a reagent	Conductance G, susceptance B, or admittance Y	High-frequency conductometric titration		
19.2.5	—	Concentration c	Relative permittivity ϵ†	Dielectrometry		The name "dielcometry" is found in the literature, but is not recommended.
19.2.6		Volume V (or otherwise measured amount) of a reagent	Relative permittivity ϵ†	Dielectrometric titration		

†Generally, though incorrectly, considered to be the dielectric constant.

TABLE 19.3 Techniques that involve double-layer phenomena but in which any electrode reactions need not be considered

Key no.	Excitation signal	Independent variable	Measured response	Recommended name of technique	Typical response curve	Remarks
19.3.1	Potential difference E	Concentration c	Interfacial tension σ between an electrode and solution (or a related parameter such as the drop time at a dropping electrode or the relative height of a polarographic maximum)	No recommendation		
19.3.2	Alternating voltage or potential E_{ac}, typically 1–5 mV	Direct potential E	Alternating current $\hat{\imath}_{ac}$ ($\hat{I}_{ac}$)	Measurement of nonfaradaic admittance		"Tensammetry," the name most widely used, is but imperfectly analogous to "voltammetry" and is therefore not recommended.

TABLE 19.4 Techniques involving electrode reactions and employing constant excitation signals

Key no.	Excitation signal (constant)	Independent variable	System	Measured response	Recommended name of technique	Typical response curve	Remarks
19.4.1	Current $i(I)$ $(=0)$	Concentration c	One indicator electrode and one reference electrode, (or two indicator electrodes) in the same solution	Potential $E = f(c)$	Potentiometry	E vs $\log c$	The terms "zero-current potentiometry" and "null-current potentiometry" are not recommended. No special terminology is recommended for measurements of pH and similar quantities.
19.4.2			Two indicator electrodes in separate solutions joined by an ionic conductor	Potential $E = f(c, c')$	Differential potentiometry	E vs $\log c$	The term "precision null-point potentiometry" is not recommended.
19.4.3		Volume V (or otherwise measured amount) of added reagent	As for potentiometry (19.4.1)	Potential $E = f(V)$	Potentiometric titration	E vs V	The terms "zero-current potentiometric titration" and "null-current potentiometric titration" are not recommended; see 19.4.10
19.4.4			As for differential potentiometry (often with one of the indicator electrodes in the titrant solution)	Potential $E = f(V)$	Differential potentiometric titration	E vs V	
19.4.5			As for potentiometry (19.4.1)	$\dfrac{dE}{dV} = f(V)$	Derivative potentiometric titration	$\dfrac{dE}{dV}$ vs V	
19.4.6				$\dfrac{dV}{dE} = f(E)$	Inverse derivative potentiometric titration	$\dfrac{dV}{dE}$ vs E	
19.4.7				$\dfrac{d^2E}{dV^2} = f(V)$	Second-derivative potentiometric titration	$\dfrac{d^2E}{dV^2}$ vs V	
19.4.8	Current $i(I)$ $(\neq 0)$	Concentration c	One indicator electrode and one reference electrode in same solution	Potential $E = f(c$ or $\log c)$	Controlled-current potentiometry	E vs $\log c$	
19.4.9			Two indicator electrodes in same solution	Potential $E = f(c$ or $\log c)$	Controlled-current potentiometry with two indicator electrodes	E vs $\log c$	The term "bipotentiometry" is no longer recommended.
19.4.10		Volume V (or otherwise measured amount) of added reagent	As for controlled-current potentiometry (19.4.8)	Potential $E = f(V)$	Controlled-current potentiometric titration	E vs V	
19.4.11			As for controlled current potentiometry with two indicator electrodes (19.4.9)	Potential $E = f(V)$	Controlled-current potentiometric titration with two indicator electrodes	E vs V	The term "bipotentiometric titration" is no longer recommended.

TABLE 19.4 (Contd)

Key no.	Excitation signal (constant)	Independent variable	System	Measured response	Recommended name of technique	Typical response curve	Remarks
19.4.12		Time t	Indicator electrode stationary in unstirred solution	Potential $E = f(t)$	Chronopotentiometry		
19.4.13				Rate of change of potential $\frac{dE}{dt} = f(t)$	Derivative chronopotentiometry		
19.4.14			Convective mass transfer to working electrode	Potential E of an indicator electrode, absorbance A, or some other composition-dependent property of the bulk of the solution electrolyzed $= f(t)$	Coulometric titration (controlled-current coulometry)		Terms like "potentiometric coulometric titration" or "controlled-current coulometry with potentiometric end-point detection" are recommended when the technique of end-point location is to be specified.
19.4.15	Applied e.m.f. or potential E	Concentration c, time t, or any other independent variable	One indicator electrode and one reference electrode	Current $i(I) = f(c)$	Amperometry		Terms like "stirred-mercury-pool amperometry" and "rotating-platinum-wire-electrode amperometry" are recommended to denote the indicator electrode employed.
19.4.16			Two indicator electrodes in same solution	Current $i(I) = f(c)$	Amperometry with two indicator electrodes		The term "biamperometry" is no longer recommended.
19.4.17			Two indicator electrodes in separate solutions joined by an ionic conductor	Current difference $\Delta i(\Delta I) = f(c, c')$	Differential amperometry		
19.4.18		Volume V (or otherwise measured amount) of added reagent	As for amperometry (19.4.15)	Current $i(I) = f(V)$	Amperometric titration		The term "amperometric titration with a dropping mercury electrode" is recommended in preference to "polarometric titration" or "polarographic titration."
19.4.19			As for amperometry with two indicator electrodes (19.4.16)	Current $i(I) = f(V)$	Amperometric titration with two indicator electrodes		The term "biamperometric titration" is no longer recommended.

TABLE 19.4 (Contd)

Key no.	Excitation signal (constant)	Independent variable	System	Measured response	Recommended name of technique	Typical response curve	Remarks
19.4.20		Time t	Stationary indicator electrode in unstirred solution	Current $i(I) = f(t)$	Chronoamperometry		The term "polarographic chronoamperometry" is recommended to denote the technique in which measurements are made during the lifetime of a single drop at a dropping electrode.
19.4.21				Quantity of electricity $Q = f(t)$	Chronocoulometry		The commonly used term "potential-step chronocoulometry" is redundant and is not recommended.
19.4.22			Convective mass transfer to working electrode	Current $i(I) = f(t)$	Convective chronoamperometry		
19.4.23			Dropping mercury (or other liquid metal) electrode as working electrode	Diffusion current $i_d(I_d) = f$ (quantity of electricity Q) or $f(t)$	Polarographic coulometry		The terms "microcoulometry" and "millicoulometry" are not recommended. The term "dropping electrode coulometry" is more specific than the recommended one and may be used when appropriate.
19.4.24			Convective mass transfer to working electrode	Mass m of material deposited on the working electrode	Electrogravimetry		The terms "internal electrogravimetry" and "spontaneous electrogravimetry" are recommended to denote spontaneous deposition.
19.4.25				Separation	Electroseparation		
19.4.26	Applied e.m.f. E or current $i(I)$		Cathodic or anodic stripping from a solid electrode into an electrolyte in a porous medium	Identification or determination of material stripped	Electrography		
19.4.27	Potential E	Time t	Convective mass transfer to working electrode	Quantity of electricity Q $\left(= \int_0^\infty i \, dt \right)$	Controlled-potential coulometry		The term "controlled-potential coulometric titration" is inappropriate and is not recommended.
19.4.28				Quantity of electricity $Q = f(t)$ or $Q_R \left[= \int_0^\infty i \, dt - \int_0^t i \, dt \right] = f(t)$	Convective chronocoulometry		

TABLE 19.4 (*Contd*)

Key no.	Excitation signal (constant)	Independent variable	System	Measured response	Recommended name of technique	Typical response curve	Remarks
19.4.29			Convective mass transfer to working electrode	Mass m of material deposited on the working electrode	Controlled-potential electrogravimetry		
19.4.30				Separation	Controlled-potential electroseparation		

TABLE 19.5 Techniques involving electrode reactions and variable excitation signals of large amplitude

Key no.	Excitation signal	Manner of variation	System	Measured response	Recommended name of technique	Typical response curve	Remarks
19.5.1	Current $i(I)$	$i = i^0 + at$	Stationary indicator electrode in unstirred solution	Potential $E = f(t)$	Chronopotentiometry with linear current sweep		
19.5.2			Dropping mercury (or other liquid conductor) electrode, or any other indicator electrode whose surface is renewed	Potential $E = f(i \text{ or } t)$	Current-scanning polarography (or polarography with linear current sweep)		
19.5.3		$i = f(t)$ (nonlinear but monotonic)	Stationary indicator electrode in unstirred solution	Potential $E = f(t)$	Programmed-current chronopotentiometry		The nature of the current-time dependence must be specified separately. For the special case in which i is a linear function of t see 5.1.
19.5.4		$i = i_1$ for $t < t_1$, $i = i_2$ for $t > t_1$	Stationary indicator electrode in unstirred solution	Potential $E = f(t)$	Current-step chronopotentiometry		The term "current-reversal chronopotentiometry" is recommended for the special case in which $i_2 = -i_1$. The term "current-cessation chronopotentiometry" is recommended for the special case in which $i_2 = 0$.
19.5.5		i periodically reversed	Stationary indicator electrode in unstirred solution	Potential E $E = f(t)$	Cyclic chronopotentiometry		The term "cyclic current-reversal chronopotentiometry" may be used to signify that $i_2 = -i_1$, and the term "cyclic current-step chronopotentiometry" may be used to signify that $i_2 \neq -i_1$. These two terms should not be used except to emphasize the difference between them.

TABLE 19.5 (Contd)

Key no.	Excitation signal	Manner of variation	System	Measured response	Recommended name of technique	Typical response curve	Remarks
19.5.6		$i = i_{ac} \sin \omega t$		Potential $E = f(t)$	Alternating-current chronopotentiometry		
19.5.7			Dropping mercury (or other liquid conductor) electrode, or any other indicator electrode whose surface is renewed	Rate of change of potential $dE/dt = f(E)$	Oscillopolarography		
19.5.8	Applied e.m.f. or potential E	$E = E^0 \pm at$	Diffusive mass transfer to any indicator electrode whose surface is not renewed	Current $i(I) = f(t)$ or, implicitly, $f(E)$	Chronoamperometry with linear potential sweep, stationary-electrode voltammetry, or linear-sweep voltammetry		Though the name "chronoamperometry with linear potential sweep" is superior in principle because all of the experimental data are dependent on the sweep rate, the alternative "stationary-electrode voltammetry" is also firmly entrenched in the literature. Single-sweep polarography (5.16) is a special case of this technique.
19.5.9			Convective mass transfer to any indicator electrode whose surface is not renewed	Current $i(I) = f(E)$	Hydrodynamic voltammetry		
19.5.10				Rate of change of current di/dt (dI/dt) or di/dE, $(dI/dE) = f(E)$	Derivative voltammetry		
19.5.11			Two indicator electrodes in separate solutions with reference electrode in each	Difference of current $dI(\Delta I) = f(E)$	Differential voltammetry		
19.5.12			Dropping mercury (or other liquid conductor) electrode, or any other indicator electrode whose surface is renewed, and a reference electrode	Current $i(I) = f(E)$	Polarography (or dc polarography)		From the viewpoint of systematic nomenclature the term "dc polarography" is a misnomer, and it should be used only when the distinction between this technique and another, such as ac polarography or rf polarography, has to be emphasized.

TABLE 19.5 (*Contd*)

Key no.	Excitation signal	Manner of variation	System	Measured response	Recommended name of technique	Typical response curve	Remarks
19.5.13				Rate of change of current $\frac{di}{dt}\left(\frac{dI}{dt}\right)$ (or, implicitly, $(di/dE)) = f(E)$	Derivative polarography (or derivative dc polarography)		See the remark about "dc polarography" above.
19.5.14			Two dropping or streaming mercury (or other liquid conductor) indicator electrodes in separate solutions, with a reference electrode in each	Difference of current $\Delta I(I) = f(E)$	Differential polarography (or differential dc polarography)		See the remark about "dc polarography" above.
19.5.15			As for polarography (5.12) but with recording of current only during the interval $t_1 \leq t \leq t_1 + \Delta t$ during the drop life, the recording device being disconnected between successive intervals	Current $i(I) = f(E)$	Tast polarography		
19.5.16			Dropping mercury (or other liquid conductor) electrode, or any other indicator electrode whose surface is renewed	Current $i(I) = f(E)$	Single-sweep polarography		In the limiting case where the sweep is so fast that the change of area during the sweep is negligible, the term "dropping-electrode chronoamperometry with linear potential (or voltage) sweep" is recommended. The terms "single-sweep oscillographic polarography" and "cathode-ray polarography" are found in the literature but are not recommended.
19.5.17			As for single-sweep polarography (19.5.16)	Current $i(I) = f(E,$ drop age)	Multisweep polarography		
19.5.18			As for single-sweep polarography (19.5.16)	Current $i(I) = f(E)$	Triangular-wave polarography		a and b may be equal or unequal.
19.5.19			As for linear-sweep voltammetry (19.5.8)	Current $i(I) = f(E)$	Triangular-wave voltammetry		
19.5.20			As for single-sweep polarography (19.5.16)	Current $i(I) = f(E)$	Cyclic triangular-wave polarography		

TABLE 19.5 (Contd)

Key no.	Excitation signal	Manner of variation	System	Measured response	Recommended name of technique	Typical response curve	Remarks
19.5.21			As for linear-sweep voltammetry (19.5.8)	Current $i(I) = f(E)$	Cyclic triangular-wave voltammetry		
19.5.22			As for single-sweep polarography (19.5.16)	Current $i(I) = f(E_1)$	Pulse polarography		It is the difference between the numbers of pulses during the life of one drop that distinguishes this technique from Kalousek polarography (19.5.24)
19.5.23		(one pulse per drop)		Difference of current Δi $(\Delta I) = f(E_1)$; implicitly $\dfrac{di}{dt}\left(\dfrac{dI}{dt}\right)$ or $\dfrac{di}{dE}\left(\dfrac{dI}{dE}\right)$ $= f(E_1)$	Derivative pulse polarography		The measured response is the difference between the direct current that flows during the interval of measurement and the direct current that flowed during the corresponding interval during the life of the preceding drop. For differential pulse polarography see 19.6.3
19.5.24		(5–50 pulses per drop)	As for single-sweep polarography (19.5.16)	Current $i(I) = f(E_1)$	Kalousek polarography		The range of variation of E_1 may be confined to values more positive than E_0, more negative than E_0, or may include E_0. Only the first possibility is shown in the illustrative curve here. Depending on the location of E_0, the range of variation of E_1, and the manner in which the recorder is connected, response curves having several different characteristic shapes may be obtained. Measurements may be made during only the latter portions of the intervals shown.
19.5.25			Indicator electrode stationary in unstirred solution	Current $i(I) = f[(t - t_1)$ and $(t - t_2)]$	Double-potential-step chronoamperometry		In this and the following technique E_1 must differ from the open-circuit potential; if it does not, the techniques are properly called "chronoamperometry" and "chronocoulometry," respectively.

TABLE 19.5 (Contd)

Key no.	Excitation signal	Manner of variation	System	Measured response	Recommended name of technique	Typical response curve	Remarks
19.5.26				Quantity of electricity $Q = f[(t-t_1)$ and $(t-t_2)]$	Double-potential-step chronocoulometry		See "double-potential-step chronoamperometry" (19.5.25)
19.5.27	Quantity of electricity Q	A charge Q', whose magnitude increases uniformly from drop to drop, is rapidly injected at drop age t'	Dropping liquid indicator electrode reference electrode	Potential $E = f(t-t')$ and slope $dE/d(t-t')^{1/2}$ of a plot of E vs $(t-t')^{1/2} = f$ (intercept of that plot)	Incremental-charge polarography		The names "charge-step polarography" and "discharge polarography" are not recommended.

TABLE 19.6 Techniques involving electrode reactions and variable excitation signals of small amplitude

Key no.	Excitation signal(s)	Manner of variation	System	Measured response	Recommended name of technique	Typical response curve	Remarks
				A: First-order techniques			
19.6.1	Current $i(I)$	$i = i_{dc} + i_{ac} \sin \omega t$	Any indicator electrode	Potential $E = f(t)$	Chronopotentiometry with superimposed alternating current		
19.6.2	Applied e.m.f. or potential	$E = E_0 + \sum_{n=0}^{0}(\Delta E)$	Dropping mercury (or other liquid conductor) electrode, or any other indicator electrode whose surface is renewed	Current $i(I) = f(E)$	Staircase polarography		n = number of steps.
19.6.3			Dropping mercury (or other liquid conductor) electrode, or any other indicator electrode surface is renewed	Difference of current $i(I) = f(E_0$ or $E_1)$	Differential pulse polarography		The measured response is the difference between the direct current that flows during the interval of measurement and the direct current that flowed during a short interval that just preceded the application of the pulse. This technique has sometimes been termed "derivative pulse polarography," but see 19.5.23
19.6.4			Dropping mercury (or other liquid conductor) electrode, or any other indicator electrode whose surface is renewed	Alternating current $i_{as}(I_{ac}) = f(E_{dc})$	Ac polarography†		The frequency of the alternating component of the applied e.m.f. or potential is usually below 1 kHz and is most often 50–60 Hz. The periodic

†The names of these techniques, which are dictated by prior usage, are among a very few that are derived from the nature of the measured response rather than from that of the excitation signal.

TABLE 19.6 (Contd)

Key no.	Excitation signal(s)	Manner of variation	System	Measured response	Recommended name of technique	Typical response curve	Remarks
							component of the excitation signal may be non-sinusoidal (e.g. triangular, sawtooth, etc.), and the technique should then be termed "polarography with superimposed periodic voltage" or, more specifically, "polarography with superimposed triangular voltage," etc. See also square-wave polarography
19.6.5		$E_{dc} = E_0 + at$ Intervals of $E_{dc} = E_0 \pm at$ measurement	Dropping mercury (or other liquid conductor) electrode, or any other indicator electrode whose surface is renewed	Square-wave current $I_{sw}(I_{sw}) = f(E_{dc})$	Square-wave polarography		This technique may be regarded as a small-amplitude analog of one variant of Kalousek polarography. It is distinguished from differential pulse polarography by involving the measurement of a periodic current rather than a direct current.
19.6.6	Current $i(I)$ and applied e.m.f. or potential E	$i = \hat{i}_{ac} \sin \omega t$ $E_{dc} = E_0 + \pm at$	Dropping mercury (or other liquid conductor) electrode, or any other indicator electrode whose surface is renewed	Alternating voltage $E_{ac} = f(E_{dc})$	Av polarography†		
19.6.7		i_{dc} = constant $E_{ac} = \hat{E}_{ac} \sin \omega t$	Dropping mercury (or other liquid conductor) electrode, or any other indicator electrode whose surface is renewed	Alternating current $i_{ac}(I_{ac}) = f(t)$	Alternating-voltage chronopotentiometry		
19.6.8				B: Second-order techniques			
	Direct applied e.m.f. or potential E_{dc} with superimposed alternating voltage E_{ac}	$E_{dc} = E_0 \pm at$ $E_{ac} = \hat{E}_{ac} \sin \omega t$	Dropping mercury (or other liquid conductor) electrode, or any other indicator electrode whose surface is renewed	Alternating current $i_{ac}(I_{ac}) = f(E_{dc})$	Higher-harmonic ac polarography		Components of the alternating current due to higher harmonics are filtered out. Typical response curves are shown for the second and third harmonics; second-harmonic ac polarography (which should be so denoted) is the most widely used of the higher-harmonic ac polarographic techniques.
19.6.9					Higher-harmonic ac polarography with phase-sensitive rectification		

†The names of these techniques, which are dictated by prior usage, are among a very few that are derived from the nature of the measured response rather than from that of the excitation signal.

TABLE 19.6 (Contd)

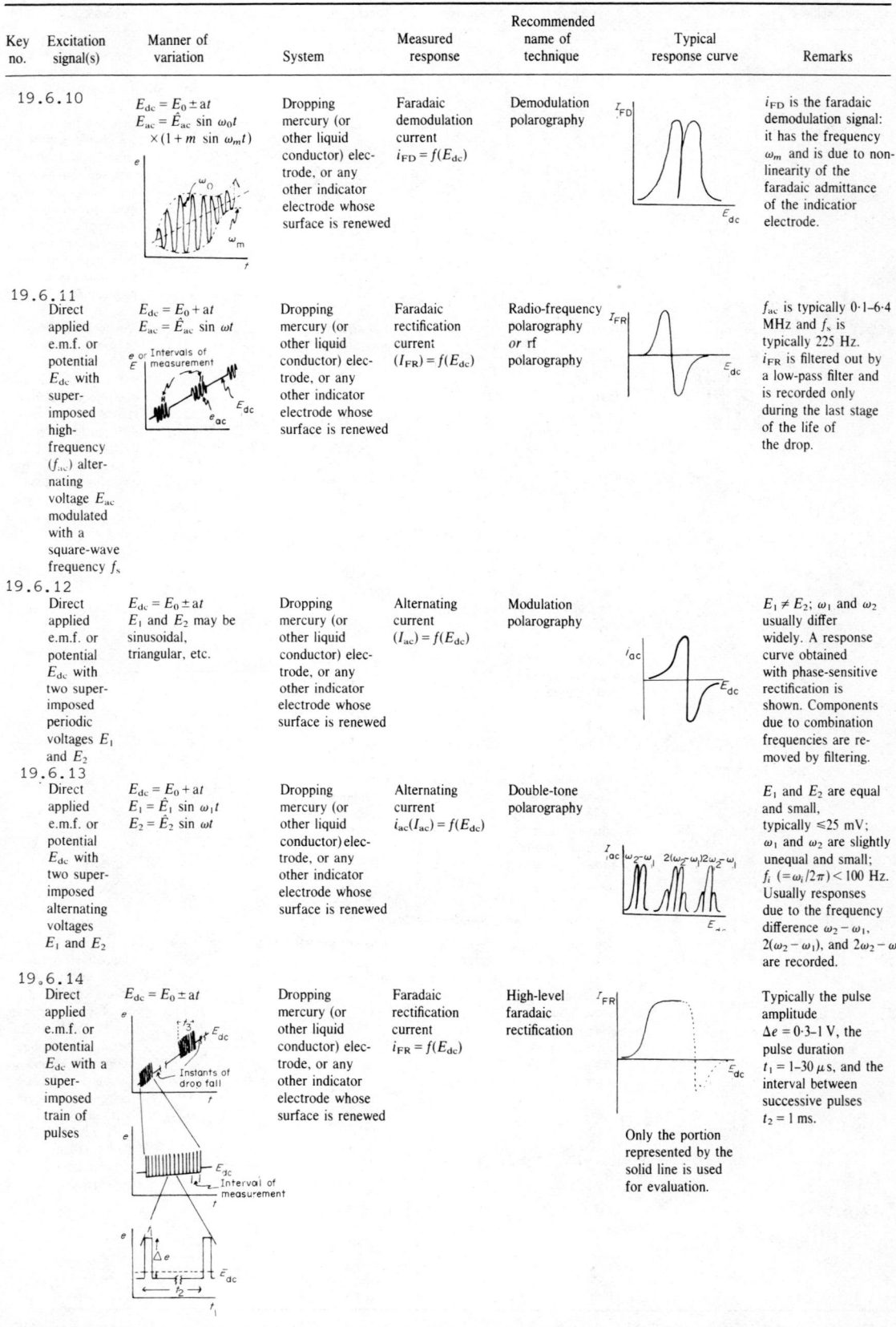

Key no.	Excitation signal(s)	Manner of variation	System	Measured response	Recommended name of technique	Typical response curve	Remarks
19.6.10		$E_{dc} = E_0 \pm at$ $E_{ac} = \hat{E}_{ac} \sin \omega_0 t$ $\times (1 + m \sin \omega_m t)$	Dropping mercury (or other liquid conductor) electrode, or any other indicator electrode whose surface is renewed	Faradaic demodulation current $i_{FD} = f(E_{dc})$	Demodulation polarography		i_{FD} is the faradaic demodulation signal: it has the frequency ω_m and is due to non-linearity of the faradaic admittance of the indicator electrode.
19.6.11	Direct applied e.m.f. or potential E_{dc} with superimposed high-frequency (f_{ac}) alternating voltage E_{ac} modulated with a square-wave frequency f_s	$E_{dc} = E_0 + at$ $E_{ac} = \hat{E}_{ac} \sin \omega t$	Dropping mercury (or other liquid conductor) electrode, or any other indicator electrode whose surface is renewed	Faradaic rectification current $(I_{FR}) = f(E_{dc})$	Radio-frequency polarography or rf polarography		f_{ac} is typically 0·1–6·4 MHz and f_s is typically 225 Hz. i_{FR} is filtered out by a low-pass filter and is recorded only during the last stage of the life of the drop.
19.6.12	Direct applied e.m.f. or potential E_{dc} with two superimposed periodic voltages E_1 and E_2	$E_{dc} = E_0 \pm at$ E_1 and E_2 may be sinusoidal, triangular, etc.	Dropping mercury (or other liquid conductor) electrode, or any other indicator electrode whose surface is renewed	Alternating current $(I_{ac}) = f(E_{dc})$	Modulation polarography		$E_1 \neq E_2$; ω_1 and ω_2 usually differ widely. A response curve obtained with phase-sensitive rectification is shown. Components due to combination frequencies are removed by filtering.
19.6.13	Direct applied e.m.f. or potential E_{dc} with two superimposed alternating voltages E_1 and E_2	$E_{dc} = E_0 + at$ $E_1 = \hat{E}_1 \sin \omega_1 t$ $E_2 = \hat{E}_2 \sin \omega t$	Dropping mercury (or other liquid conductor) electrode, or any other indicator electrode whose surface is renewed	Alternating current $i_{ac}(I_{ac}) = f(E_{dc})$	Double-tone polarography		E_1 and E_2 are equal and small, typically ≤25 mV; ω_1 and ω_2 are slightly unequal and small; $f_i (=\omega_i/2\pi) < 100$ Hz. Usually responses due to the frequency difference $\omega_2 - \omega_1$, $2(\omega_2 - \omega_1)$, and $2\omega_2 - \omega_1$ are recorded.
19.6.14	Direct applied e.m.f. or potential E_{dc} with a superimposed train of pulses	$E_{dc} = E_0 \pm at$	Dropping mercury (or other liquid conductor) electrode, or any other indicator electrode whose surface is renewed	Faradaic rectification current $i_{FR} = f(E_{dc})$	High-level faradaic rectification	Only the portion represented by the solid line is used for evaluation.	Typically the pulse amplitude $\Delta e = 0.3$–1 V, the pulse duration $t_1 = 1$–30 μs, and the interval between successive pulses $t_2 = 1$ ms.

TABLE 19.7. Index to Tables 19.2 - 19.6

Ac polarography, 19.6.4
  higher-harmonic, 19.6.8
    with phase-sensitive rectification, 19.6.9
Alternating current chrono-
  potentiometry, 19.5.6
Alternating-voltage chrono-
  potentiometry, 19.6.7
Amperometric titration, 19.4.18
  with two indicator electrodes, 19.4.19
Amperometry, 19.4.15
  differential, 19.4.17
  stirred-mercury-pool, 19.4.15
  with two indicator electrodes, 19.4.16
Av polarography, 19.6.6
Biamperometric titration, 19.4.19
Biamperometry, 19.4.16
Bipotentiometric titrations, 19.4.11
Bipotentiometry, 19.4.9
Cathode-ray polarography, 19.5.16
Charge-step polarography, 19.5.27
Chronoamperometry, 19.4.20
  convective, 19.4.22
  double-potential-step, 19.5.25
  dropping electrode with linear potential,(or voltage) sweep, 19.5.16
  polarographic, 19.4.20
  stirred-mercury-pool, 19.4.22
  with linear potential sweep, 19.5.8
Chronocoulometry, 19.4.21
  convective, 19.4.28
  double-potential-step, 19.5.26
  potential-step, 19.4.21
Chronopotentiometry, 19.4.21
  alternating-current, 19.5.6
  alternating-voltage, 19.6.7
  current-cessation, 19.5.4
  current reversal, 19.5.4
  current step, 19.5.4
  cyclic, 19.5.4
  cyclic current-reversal, 19.5.5
  derivative, 19.4.13
  programmed-current, 19.5.3
  with linear current sweep, 19.5.1
  with superimposed alternating current, 19.6.1

Conductimetric titration, 19.2.2
  high-frequency, 19.2.4
Conductometry, 19.2.1
  high-frequency, 19.2.3
Controlled-current coulometry, 19.4.14
Controlled-current potentiometric titration, 19.4.10
Controlled-current potentiometry, 19.4.8
Controlled-potential coulometric titration, 19.4.27
Controlled-potential coulometry, 19.4.27
Controlled-potential electrogravimetry, 19.4.29
Controlled-potential electroseparation, 19.4.30
Convective chronoamperometry, 19.4.22
Convective chronocoulometry, 19.4.28
Coulometric titration, 19.4.14
  controlled-potential, 19.4.27
Coulometry, controlled-current, 19.4.14
  controlled-potential, 19.4.27
  dropping-electrode, 19.4.23
  polarographic, 19.4.23
Current-cessation chrono-
  potentiometry, 19.5.4
Current-reversal chrono-
  potentiometry, 19.5.4
  cyclic, 19.5.5
Current-scanning polarography, 19.5.2
Current-step chronopotentiometry, 19.5.5
Cyclic chronopotentiometry, 19.5.5
Cyclic triangular-wave polarography, 19.5.20
Dc conductometry, 19.2.1
Dc polarography, 19.5.12
Demodulation polarography, 19.6.10
Derivative chronopotentiometry, 19.4.13
Derivative polarography, 19.5.13
Derivative potentiometric titration, 19.4.5
  inverse, 19.4.6
Derivative pulse polarography, 19.5.23
Dielcometric titration, 19.2.6
Dielcometry, 19.2.5
Dielectrometric titration, 19.2.6
Dielectrometry, 19.2.5
Differential amperometry, 19.4.17
Differential potentiometric titration, 19.4.4
Differential potentiometry, 19.4.2
Differential pulse polarography, 19.6.3

Differential voltammetry, 19.5.11
Discharge polarography, 19.5.27
Double-potential-step chrono-
potentiometry, 19
Double-potential-step chrono-
coulometry, 19.5.25
Double-tone polarography, 19.6.13
Dropping-electrode coulometry, 19.4.23
Electrography, 19.4.26
Electrogravimetry, 19.4.24
 controlled potential, 19.4.29
Electroseparation, 19.4.25
 controlled potential, 19.4.30
Faradaic rectification, high-level, 19.6.14
High-frequency conductometric
titration, 19.2.4
High-frequency conductometry, 19.2.3
High-level faradaic rectification, 19.6.14
Higher-harmonic ac polarography, 19.6.8
 with phase-sensitive rectification, 19.6.9
Hydrodynamic voltammetry, 19.5.9
Incremental-charge polarography, 19.5.27
Interfacial tension, measurement of, 19.3.1
Internal electrogravimetry, 19.4.24
Inverse derivative potentiometric
titration, 19.4.6
Linear-sweep voltammetry, 19.5.8
Kalousek polarography, 19.5.24
Microcoulometry, 19.4.23
Millicoulometry, 19.4.23
Modulation polarography, 19.6.12
Multisweep polarography, 19.5.17
Non-faradaic admittance,
measurement of, 19.3.2
Null-current potentiometric titration, 19.4.3
Null-current potentiometry, 19.4.1
Oscillopolarography, 19.5.7
Oscillographic polarography, 19.5.7
 single sweep, 19.5.16
Polarographic chronoamperometry, 19.4.20
Polarographic coulometry, 19.4.23
Polarographic titration, 19.4.18
Polarography, 19.5.12
 ac, 19.6.4
  higher-harmonic, 19.6.8
   with phase-sensitive rectification, 19.6.9
  av, 19.6.6

Polarography,
 cathode-ray, 19.5.16
 charge-step, 19.5.27
 current-scanning, 19.5.2
 cyclic triangular-wave, 19.5.20
 demodulation, 19.6.10
 derivative, 19.5.13
 derivative pulse, 19.5.23
 differential, 19.5.14
 differential pulse, 19.6.3
 discharge, 19.5.29
 double-tone, 19.6.13
 higher-harmonic, 19.6.8
  ac, with phase-sensitive
   rectification, 19.6.9
 incremental-charge, 19.5.27
 Kalousek, 19.5.24
 modulation, 19.6.12
 multisweep, 19.5.17
 oscillographic, 19.5.7
 pulse, 19.5.22
  derivative, 19.5.23
  differential, 19.6.3
 radiofrequency (rf), 19.6.11
 single-sweep, 19.5.16
  oscillographic, 19.5.16
 square-wave, 19.6.5
 staircase, 19.6.2
 Tast, 19.5.15
 triangular-wave, 19.5.18
  cyclic, 19.5.20
Polarometric titration, 19.4.18
Potential-step chronocoulometry, 19.4.21
(note, 19.5.26)
Potentiometric titration, 19.4.3
 controlled-current, 19.4.10
 derivative, 19.4.5
 differential, 19.4.4
 inverse derivative, 19.4.6
 null-current, 19.4.3
 second-derivative, 19.4.7
 with two indicator electrodes, 19.4.11
 zero-current, 19.4.3
Potentiometry, 19.4.1
 controlled current, 19.4.8
 differential, 19.4.2
 null-current, 19.4.1
 precision null-point, 19.4.2

Potentiometry,
  with two indicator electrodes, 19.4.9
  zero-current, 19.4.1
Precision null-point potentiometry, 19.4.2
Programmed-current chrono-
potentiometry, 19.5.3
Pulse polarography, 19.5.22
  derivative, 19.5.23
  differential, 19.6.3
Radiofrequency (rf) polarography, 19.6.11
Second-derivative potentiometric
titration, 19.4.7
Single-sweep polarography, 19.5.16
Single-sweep oscillographic
polarography, 19.5.16
Spontaneous electrogravimetry, 19.4.24
Square-wave polarography, 19.6.5
Staircase polarography, 19.6.2
Stirred-mercury-pool amperometry, 19.4.15
Tast polarography, 19.5.15
Tensammetry, 19.3.2
Titration, amperometric, 19.4.18
  with two indicator electrodes, 19.4.19
  biamperometric, 19.4.19
  bipotentiometric, 19.4.11
  conductometric, 19.2.2
    high-frequency, 19.2.4
  controlled-current potentiometric, 19.5.10
  controlled-current coulometric, 19.4.27
  coulometric, 19.4.14
    controlled-potential, 19.4.27
  derivative potentiometric, 19.4.5
  dielcometric, 19.2.6
  dielectrometric, 19 2.6

Titration,
  differential potentiometric, 19.4.4
  high-frequency conductometric, 19.2.4
  inverse derivative potentiometric, 19.4.6
  null-current potentiometric, 19.4.3
  polarographic, 19.4.18
  polarometric, 19.4.18
  potentiometric, 19.4.3
    controlled-current, 19.4.10
    derivative, 19.4.5
    differential, 19.4.4
    inverse derivative, 19.4.6
    null-current, 19.4.3
    second-derivative, 19.4.7
    with two indicator electrodes, 19.4.11
    zero-current, 19.4.3
  second-derivative potentiometric, 19.4.7
  zero-current potentiometric, 19.4.3
Triangular-wave polarography, 19.5.18
  cyclic, 19.5.20
Triangular-wave voltammetry, 19.5.19
  cyclic, 19.5.21
Voltammetry, 19.5.9
  cyclic triangular-wave, 19.5.21
  derivative, 19.5.10
  differential, 19.5.11
  hydrodynamic, 19.5.9
  linear-sweep, 19.5.8
  stationary-electrode, 19.5.8
  triangular-wave, 19.5.19
    cyclic, 19.5.21
Zero-current potentiometric
titration, 19.4.3
Zero-current potentiometry, 19.4.1

20. RECOMMENDATIONS FOR SIGN CONVENTIONS AND PLOTTING OF ELECTROCHEMICAL DATA*

The Commission on Electroanalytical Chemistry of the International Union of Pure and Applied Chemistry wishes to alert authors to an impending agreement on a sign convention for currents, which is at variance with prevailing practice in the electroanalytical literature. For the sake of uniformity, we urge electroanalytical chemists to use this IUPAC sign convention, which is outlined below.

The fundamental convention will consist of assigning positive values to anodic currents and negative ones to cathodic currents. Anodic and cathodic currents will continue to be defined as corresponding to net oxidation and net reduction, respectively, at the indicator or working electrode.

Conformity to this convention will require many chemists who work with polarographic waves, chronopotentiograms, and other electrochemical response curves to reformulate some of the equations associated with them and to adjust related procedures.

Any reasonable choice of coordinates is appropriate in plotting any such curve, provided that the abscissa and ordinate axes are clearly labelled. Most of the polarographic and other voltammetric curves in the existing literature are plotted with cathodic currents above the abscissa axis and negative values of the applied e.m.f. to the right of the ordinate axis. Those who wish to follow the new convention and also to facilitate comparison of their curves with those in the prior literature may achieve both aims by choosing $-i$ † as the positive ordinate and $-E$ as the positive abscissa.

Some relationships and plots will be unaltered by the adoption of the new convention. An example is the analysis of a polarographic wave obtained for a Nernstian ('reversible') diffusion-controlled half-reaction of the type

$$M^{n+} + ne + Hg = M(Hg)$$

Obviously, a line having a slope of $2.3RT/nF$ volt for the cathodic process will still be obtained if $-E$ is plotted against $\log_{10}[i/(i_{d,c} - i)]$, because the argument of the logarithmic term will still be positive at every point even though i (the cathodic current at the potential E) and $i_{d,c}$ (the cathodic diffusion current) are both regarded as negative. The corresponding plot for a composite anodic-cathodic wave would be one of $-E$ against

---

† The symbol I for electric current is recommended by IUPAC and other international organizations. However, the symbol i, generally used in the electroanalytical literature, is acceptable.

* Based upon the approved Recommendations published in *Pure and Applied Chemistry*, Vol. 45, No.2 (1976), pp. 131 - 134.

$\log_{10}[(i - i_{d,a})/(i_{d,c} - i)]$, where $i_{d,a}$ is the anodic diffusion current, now regarded as positive, and this again conforms to current practice. It may be stressed that the familiar form is retained because the new convention alters the signs of both $(i - i_{d,a})$ and $(i_{d,c} - i)$.

Such cancellations of sign occurs only when the ratio of two currents is involved, and expressions that involve only a single current will require adjustment. For instance, the equation of a Nernstian ('reversible') polarographic wave corresponding to the process

$$M^{n+} + ne = M(\text{insoluble})$$

of which one common form is

$$E = \text{constant} + \frac{RT}{nF}\ln(i_{d,c} - i)$$

will have to be rewritten as

$$E = \text{constant} + \frac{RT}{nF}\ln(i - i_{d,c})$$

to avoid assigning negative values to the argument of the logarithmic term. The cathodic current i has generally been related to the difference between the concentration c of an electroreducible substance in the bulk of a solution and its concentration c^o at the surface of the indicator electrode by equations of the forms

$$i = k(c - c^o) \text{ and } i = nFAk_{red}c^o$$

for diffusion- and rate-controlled processes, respectively, and these will have to be rewritten as

$$i = -k(c - c^o) \text{ and } i = -nFAk_{red}c^0$$

so that the cathode current will always have the prescribed negative sign.

Other electrochemical equations should be modified, by introducing or removing a minus sign, in the fashion illustrated by the last two examples. For instance, the Ilkovič and Sand equations should be written as

$$i_d = -knD^{1/2}c_m^{2/3}\tau^{1/6}$$

and

$$i = -\pi^{1/2}nFAD^{1/2}c/2\tau^{1/2}$$

Polarographic diffusion current constants, chronopotentiometric constants, and similar quantities should continue to have the same signs as the currents to which they pertain, and should accordingly be taken as negative for cathodic processes.

It is strongly urged that due consideration be given to all relevant IUPAC conventions and to problems of internal consistency in using all equations or definitions likely to be affected by the new convention, and in specifying the sign of any quantity appearing in those equations.

21. RECOMMENDATIONS FOR NOMENCLATURE OF ION-SELECTIVE ELECTRODES*

21.1 GENERAL RECOMMENDATIONS

21.1.01 DEFINITION OF TERMS. For definitions of the terms *activity, activity coefficient and concentration* refer to the *Manual of Symbols and Terminology for Physicochemical Quantities and Units* (Butterworths, London, 1973 Edition).

21.1.02 CALIBRATION CURVE. This is a plot of the potential (emf) of a given ion-selective electrode cell assembly (ion-selective electrode combined with an identified reference electrode) versus the logarithm of the ionic activity (concentration) of a given species. For uniformity, it is recommended that the potential be plotted on the ordinate (vertical axis) with the more positive potentials at the top of the graph, and that pa_A (- log activity of the species measured, A) or pc_A be plotted on the abscissa (horizontal axis) with increasing activity to the right.

21.1.03 LIMIT OF DETECTION. A calibration curve ordinarily has the shape shown in Fig. 21.1.

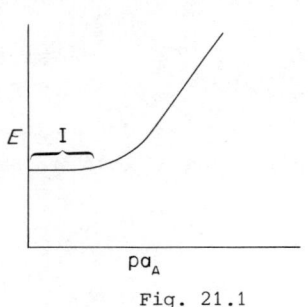

Fig. 21.1

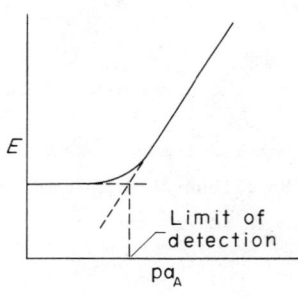

Fig. 21.2

By analogy with definitions adopted in other fields, the limit of detection should be defined as the concentration for which, under specified conditions, the potential E deviates from the average potential in region I by some arbitrary multiple of the standard error of a single measurement of the potential in region I.

In the present state of the art, and for the sake of practical convenience, a simpler (and more convenient definition is recommended at this time. The practical limit of detection may be taken as the activity (or concentration) of A at the point of intersection of the extrapolated segments of the calibration curve, as shown in Fig. 21.2.

---

* Based on the approved Recommendations (1975) published in *Pure and Applied Chemistry*, Vol.48 (1976), pp. 127 - 132.

Since many factors affect the detection limit, the experimental conditions used should be reported, i.e. the composition of the solution, the history and preconditioning of the electrode, stirring, etc.

21.1.04 DRIFT. This is the slow non-random change with time in the potential (emf) of an ion-selective electrode cell assembly maintained in a solution of constant composition and temperature.

21.1.05 HYSTERESIS (ELECTRODE MEMORY). Hysteresis is said to have occurred if, after the concentration has been changed and then restored to its original value, there is a different potential observed. The reproducibility of the electrode will consequently be poor. The systematic error is generally in the direction of the concentration of the solution in which the electrode was previously immersed.

21.1.06 MEMBRANE. This refers to a continuous layer covering a structure or separating two electrolytic solutions. The membrane of an ion-selective electrode is responsible for the potential response and selectivity of the electrode (see 21.2 for listing of membranes).

21.1.07. ION-SELECTIVE ELECTRODES. These are electrochemical sensors, the potentials of which are linearly dependent on the logarithm of the activity of a given ion in solution. Such devices are distinct from systems which involve redox reactions (Class I and II Electrodes).

Comment: The term "ion-specific electrode" is not recommended. The term "specific" implies that the electrode does not respond to additional ions. Since no electrode is truly specific for one ion, the term "ion-selective" is recommended as more appropriate. "Selective ion-sensitive electrode" is a little-used term to describe an ion-selective electrode.

The potential response has as its principal component the free-energy change associated with mass transfer (by ion-exchange, adsorption, liquid-liquid extraction, or some other mechanism) across a phase boundary.

21.1.08 INTERFERING SUBSTANCE. This is any species, other than the ion being measured, whose presence in the sample solution affects the measured potential of the cell.

Interfering substances fall into two classes: "electrode" interferences and "method" interferences. Examples of the first class would be those substances which give a similar response to the ion being measured and whose presence generally results in an apparent increase in the activity (or concentration) of the ion to be determined (e.g. Na^+ for the Ca^{2+} electrode), those species which interact with the membrane so as to change its chemical composition (e.g. organic solvents for the liquid or polyvinylchloride (PVC) membrane electrodes), or electrolytes present at a high concentration giving rise to appreciable liquid-junction potentials. The second class of interfering substance is that which interacts with the ion being measured so as to decrease its activity or apparent concentration, but where the electrode continues to report the true activity (e.g. CN^- present in the measurement of Ag^+.

21.1.09 REFERENCE ELECTRODE. An electrode which maintains a virtually invariant potential under the conditions prevailing in an electrochemical measurement, and which serves to permit the observation or control the potential of the indicator (or test) or working electrode.

Comment: Practical reference electrodes are generally constructed so that their electrolyte solutions serve as salt bridges to the solutions under investigation.

21.1.10 INTERNAL REFERENCE ELECTRODE. This is a reference electrode which is contained inside an ion-selective assembly.

Comment: The system frequently consists of a silver - silver chloride electrode in contact with an appropriate solution containing chloride and a fixed concentration of the ion for which the membrane is selective.

21.1.11 IONIC-STRENGTH ADJUSTMENT BUFFER. A pH buffered solution of high ionic strength added to samples and calibration solutions before measurement in order to achieve identical ionic strength and hydrogen ion activity. In addition complexing agents and other components are often added to minimise the effects of certain interferences.

21.1.12 NERNSTIAN RESPONSE. An ion-selective electrode is said to have a Nernstian response over a given range of activity (or concentration) in which a plot of the potential of such an electrode in conjunction with a reference electrode versus the logarithm of the ionic activity of a given species (a_A) is linear with a slope of $2.303 \times 10^3 RT/z_A F$ mV/decade ($59.16/z_A$ mV per unit of pa_A at 25°C).

21.1.13 PRACTICAL RESPONSE TIME. The length of time which elapses between the instant at which an ion-selective electrode and a reference electrode are brought into contact with a sample solution (or at which the concentration of the ion of interest in a solution in contact with an ion-selective electrode and a reference electrode is changed) and the first instant at which the potential of the cell becomes equal to its steady-state value within 1 mV. The experimental conditions used should be stated, i.e. the stirring rate, the composition of the solution of which the response time is measured, the composition of the solution to which the electrode was exposed prior to this measurement, the history and the preconditioning of the electrode, and the temperature.

21.1.14 COMBINATION ELECTRODE. An electrochemical apparatus which incorporates an ion-selective electrode and a reference electrode in a single assembly, thereby avoiding the need for a separate reference electrode.

21.1.15 POTENTIOMETRIC SELECTIVITY COEFFICIENT, $k_{A,B}^{pot}$. This defines the ability of an ion-selective electrode to distinguish between different ions in the same solution. It is not identical to a similar term used in separation processes. The selectivity coefficient is evaluated by means of the ion-selective electrode emf response, in mixed solutions of the primary ion, A, and interfering ion, B, (or less desirably, in separate solutions). The activities of the primary ion, A, and the interfering ion, B, at which $k_{A,B}^{pot}$ is determined should always be specified, as the value of $k_{A,B}^{pot}$ is defined by the modified Nernst equation. The smaller the value of $k_{A,B}^{pot}$, the greater the electrode's preference for the principal ion, A, as described later.

Comment: The terms *selectivity constant* and *selectivity factor* are frequently used instead of selectivity coefficient. However, in order to standarize the terminology associated with ion-selective electrodes, use of the term selectivity coefficient is now recommended, as is the fixed interference method for its evaluation (see 21.3.4.2).

21.1.16 STANDARD ADDITION OR KNOWN ADDITION METHOD. This is a procedure for the determination of the concentration of a particular species in a sample by adding known amounts of that species to the sample solution and recording the change of potential of an ion-selective electrode versus a suitable reference electrode.

21.1.17 STANDARD SUBTRACTION OR KNOWN SUBTRACTION. This is a variation of the standard addition method. In this procedure changes in the potential resulting from the addition of a known amount of a species which reacts stoichiometrically with the ion of interest (e.g. a complexing agent) are employed to determine the original activity or concentration of the ion.

21.1.18 ISOPOTENTIAL POINT. For a cell containing an ion-selective electrode and a reference electrode there is often a particular activity of the ion concerned for which the potential of the cell is independent of temperature. That activity, and the corresponding potential, define the isopotential point. The identity of the reference electrode, and the composition of the filling solution of the measuring electrode, must be specified.

21.2 CLASSIFICATION OF ION-SELECTIVE ELECTRODES

21.2.1 *Primary electrodes*

21.2.1.1 CRYSTALLINE ELECTRODES. These may be homogeneous or heterogeneous.

21.2.1.1.1 HOMOGENEOUS MEMBRANE ELECTRODES are ion-selective electrodes in which the membrane is a crystalline material prepared from either a single compound or from a homogeneous mixture of compounds (i.e., Ag_2S, AgI/Ag_2S).

21.2.1.1.2 HETEROGENEOUS MEMBRANE ELECTRODES formed when an active substance, or mixture of active substances, is mixed with an inert matrix, such as silicone rubber or PVC, or placed on hydrophobized graphite, to form the sensing membrane which is heterogeneous in nature.

21.2.1.2 NON-CRYSTALLINE ELECTRODES. In these electrodes a support, containing an ionic (either cationic or anionic) species or an uncharged species, forms the ion-selective membrane which is usually interposed between two aqueous solutions. The support can either be porous (e.g. Millipore filter, glass frit, etc.) or non-porous (e.g. glass or inert polymeric material such as PVC, yielding with the ion-exchanger and the solvent a "solidified" homogeneous mixture). These electrodes exhibit a response due to the presence of the ion-exchange material in the membrane.

21.2.1.3 RIGID MATRIX ELECTRODES (e.g. GLASS ELECTRODES). These are ion-selective electrodes in which the sensing membrane is a thin piece of glass. The chemical composition of the glass determines the selectivity of the membrane. In this group are:

   hydrogen ion-selective electrodes

   monovalent cation-selective electrodes.

21.2.2 *Electrodes with a mobile carrier*

21.2.2.1 POSITIVELY CHARGED — bulky cations (e.g. those of quaternary ammonium salts or salts of transition metal complexes such as derivatives of 1,10-phenanthroline) which, when dissolved in a suitable organic solvent and held on an inert support (e.g. a Millipore filter or PVC), provide membranes which are sensitive to changes in the activities of anions.

21.2.2.2 NEGATIVELY CHARGED —Complexing agents (e.g. of the type $(RO)_2PO_2^-$) or bulky anions (e.g. tetra-*p*-chlorophenylborate anions) which, when dissolved in a suitable organic solvent and held in an inert support (e.g. a Millipore filter or PVC), provide membranes that are sensitive to changes in the activities of cations.

21.2.2.3 UNCHARGED CARRIER — Electrodes based on solutions of molecular carriers of cations (e.g. antibiotics, macrocyclic compounds or other sequestering agents) which can be used in membrane preparations which show sensitivity and selectivity to certain cations.

21.2.3 *Sensitized ion-selective electrodes*

21.2.3.1 GAS SENSING ELECTRODES. These are sensors composed of an indicating and a reference electrode which use a *gas-permeable membrane* or an *air-gap* to separate the sample solution from a thin film of an intermediate solution, which is either held between the gas membrane and the ion-sensing membrane of the electrode, or placed on the surface of the electrode using a wetting agent (e.g. air-gap electrode). This intermediate solution interacts with the gaseous species in such a way as to produce a change in a measured value (e.g. pH) of the intermediate solution. This change is then sensed by the ion-selective electrode and is proportional to the partial pressure of the gaseous species in the sample.

Note: An exception to this classification is the hydrogen gas electrode which responds both to the partial pressure of hydrogen and to pH. The oxygen electrode fits under this classification although, in contrast to all other sensors, it is an amperometric and *not* a potentiometric device.

21.2.3.2 ENZYME SUBSTRATE ELECTRODES are sensors in which an ion-selective electrode is covered with a coating containing an enzyme which causes the reaction of an organic or inorganic substance (substrate) to produce a species to which the electrode responds. Alternatively, the sensor could be covered with a layer of substrate which reacts with the enzyme to be assayed.

21.3 CONSTANTS AND SYMBOLS

21.3.1 THE MODIFIED NERNST EQUATION FOR ION-SELECTIVE ELECTRODES AND DEFINITION OF $k^{pot}_{A,B}$

$$E = \text{constant} + \frac{2.303RT}{z_A F} \log(a_A + k^{pot}_{A,B}(a_B)^{z_A/z_B} + k^{pot}_{A,C}(a_C)^{z_A/z_C} \ldots)$$

E is the experimentally observed potential of a cell (in millivolts)

R is the gas constant and is equal to 8.31441 J K^{-1}mol^{-1}

T is the thermodynamic temperature (in K)

F is the Faraday constant and is equal to $(9.648670 \pm 0.000054) \times 10^4$ C mol^{-1}

a_A is the activity of the ion, A

a_B and a_C are the activities of the interfering ions, B and C, respectively

$k^{pot}_{A,B}$ is the potentiometric selectivity coefficient

z_A is an integer with sign and magnitude corresponding to the charge of the principal ion, A

z_B and z_C are integers with sign and magnitude corresponding to the charges of the interfering ions, B and C, respectively.

The "constant" term includes the standard or zero potential of the indicator electrode, E^o_{ISE}, the reference electrode potential, E_{Ref}, and the junction potential, E_j (all in millivolts).

21.3.2 IONIC STRENGTH OF A SOLUTION. This is defined by $I = \frac{1}{2}\Sigma c_i z_i^2$ where I is the ionic strength, c_i is the concentration in mole per litre of an ion, i, and z_i is the charge on the ion, i.

21.3.3 OTHER SYMBOLS. Sign conventions should be in accord with IUPAC recommendations (*Manual of Symbols and Terminology for Physicochemical Quantities and Units*, Butterworths, London, 1973 edition, p.27).

21.3.4 METHODS FOR DETERMINING $k_{A,B}^{pot}$.

21.3.4.1 FIXED INTERFERENCE METHOD. The potential of a cell comprising an ion-selective electrode and a reference electrode is measured with solutions of constant level of interference, a_B, and varying activity of the primary ion, a_A. The potential values obtained are plotted against the activity of the primary ion. The intersection of the extrapolation of the linear portions of this curve (see Fig. 21.2) will indicate the values of a_A which are to be used to calculate $k_{A,B}^{pot}$ from the equation

$$k_{A,B}^{pot} = a_A / (a_B)^{z_A/z_B}$$

21.3.4.2 SEPARATE SOLUTION METHOD. The potential of a cell comprising an ion-selective electrode and a reference electrode is measured with each of two separate solutions, one containing the ion A at the activity a_A (but no B), the other containing the ion B at the same activity $a_B = a_A$ (but no A). If the measured values are E_1 and E_2, respectively, the value of $k_{A,B}^{pot}$ may be calculated from the equation

$$\log k_{A,B}^{pot} = \frac{E_2 - E_1}{2.303 RT/z_A F} + (1 - \frac{z_A}{z_B}) \log a_A$$

The method is only recommended if the electrode exhibits a Nernstian response. It is less desirable because it does not represent as well the actual conditions under which the electrodes are used.

APPENDIX
RECOMMENDATIONS ON THE USAGE OF
THE TERMS 'EQUIVALENT' AND 'NORMAL'

INTRODUCTION

This appendix has been included as an addition to the original manuscript following the approval of both by the Council at its 29th General Assembly in Warsaw, August 1977. The original version of the report on 'Equivalent' and 'Normal' is being published independently in *Pure and Applied Chemistry*, Volume 50 (1978).

Following the introduction of the SI and the IUPAC *Manual of Symbols and Terminology for Physicochemical Quantities and Units* it has become necessary to clarify the conditions under which the traditional and technologically convenient 'Equivalence' and 'Normal' concepts and terms should now be used for quantitative analytical work in aqueous solutions. The argument is restricted to the consideration of acid-base and redox reactions in aqueous solutions since this is the only area in which these terms are widely used.

The present report does not imply any recommendation that the terms 'Equivalent' and 'Normal' should continue to be used. It does, however, provide the necessary guidelines for those who are required to or may still wish to use such terms.

THE CONCEPT OF EQUIVALENCE AND NORMAL SOLUTIONS

The concept of equivalence between the amounts of reacting substances has played a fundamental part in the history of quantitative chemistry and its development as an exact science[1]. Its role in titrimetric analysis is equally fundamental and scarcely needs stressing. If, for example, we consider a basic type of reaction

$$\nu_A A + \nu_B B \longrightarrow \text{products} \tag{1}$$

between a species A (the analyte) in one solution (the sample solution) and a species B which reacts with it stoicheiometrically and is contained in a second solution* (the titrating solution or titrant) the molar masses of the two species which are equivalent are $\nu_A M_A$ and $\nu_B M_B$ where M_A and M_B are the molar masses of the two species (formerly called the gram molecular weights), and ν_A and ν_B are the respective number of reacting

---

* In many early procedures the titrant was added as a solid to a solution of the analyte though the reverse of this procedure was also common.

entities (now termed the stoicheiometric number of the components)[2].

A development of profound importance in practical analysis was the realisation that titrimetric procedures could be carried out with greater speed and convenience if the concentrations of the two reacting solutions were such that the reaction with the analyte was complete when comparable *volumes* of sample and titrant solutions had been brought together. More specifically, if volumes V_A and V_B of these solutions were mixed the reaction would be stoicheiometric when $N_A V_A = N_B V_B$ where 'N_X' the 'normality' of the solution designated the number of 'gram equivalents' per litre.

Since it has become clear[3] that there is still a general desire among those who use titrimetric procedures extensively to continue to use much of this convenient terminology it becomes essential to re-examine the nomenclature to make sure that terms such as 'normal' and 'equivalent' should be clearly defined and that any units employed must be those approved by *Le Système International d'Unités* (SI); furthermore, there should be no inconsistencies with established IUPAC recommendations already approved and published[3,4].

The concept of *equivalence* and the use of the term *equivalent* is well established in studies of ion-exchange phenomena and in electroanalytical chemistry (notably in electrogravimetry and conductimetric procedures) and any proposals made for standardisation of terminology in titrimetric analysis must, of course, be equally applicable to these and other relevant fields.

THE SI SYSTEM AND ITS IMPLICATION

The international adoption of SI, especially the new base unit for *amount of substance* (the mole), has meant that a number of terms widely used in analytical chemistry are no longer really necessary. Some are undoubtedly being used in senses which are not in accordance with and even conflict with the precise use of SI.

The vast majority of chemists who received their education before the early 1970s have regarded and may still regard the expression 'one mole of NaOH' to be defined as a definite mass (weight) of this compound, *i.e.* 'one gram molecular weight', 40 g of sodium hydroxide.

The term 'mole' is now used in a more precise sense, and before further progress can be made it seems essential to restate some basic ideas to establish unambiguously the relationship between the current and older terminologies. Only then will it be possible to formulate proposals covering the use of such concepts as 'equivalence' and 'normality'.

THE AMOUNT OF SUBSTANCE

The SI base unit for 'amount of substance' is the *mole* defined as follows. "The mole is the amount of substance of a system which contains as many elementary entities as there are atoms in 0.012 kilogram of carbon-12. When the mole is used the elementary entities must be specified and may be atoms, molecules, ions, electrons, other particles or specified groups of such particles".

The symbol for amount of substance is n; the amount of substance of species X is symbolised $n(X)$.

USAGE OF TERMS 'EQUIVALENT' AND 'NORMAL'

Examples
$n(Mg^{2+})$ = 5 mmol
$n(KMnO_4)$ = 0.1 mol
$n(F)$ = 6 mmol
$n(C_2H_5OH)$ = 1 kmol
$n(H^+)$ = 10^{-6} mol

Knowing values for A_r, the *relative atomic mass* (formerly the 'atomic weight') from appropriate Tables for each constituent element, we can calculate the corresponding mass.

Examples
5 mmol of Mg^{2+} has a mass of 0.012116 g
0.1 mol of $KMnO_4$ has a mass of 15.804 g
6 mmol of F has a mass of 0.114 g
1 kmol of C_2H_5OH has a mass of 46.070 kg

It must be emphasized that the mole concept refers to any specified particle or group of particles and we can speak of electrons

$n(e^-)$ = 1 mol (with a mass of 0.548 6 x 10^{-3} g)

or of a doped crystal of specified composition, *e.g.*

$n(Na_{0.93}Tl_{0.07})$ = 0.2 mol (with a mass of 7.138 g)

Although the SI base unit of mass is the kilogram, decimal multiples and submultiples are, of course, acceptable.

MOLAR MASS   The molar mass (symbol M) is defined as mass divided by amount of substance. The SI base unit is kg mol^{-1} and the practical unit is usually g mol^{-1}.

Examples
$M(Cu)$ = 63.54 g mol^{-1}
$M(H^+)$ = 1.0074 g mol^{-1}
$M(Cl_2)$ = 70.916 g mol^{-1}

RELATIVE ATOMIC MASS

The *relative atomic mass*, A_r (formerly called the 'atomic weight'), is the average mass per atom of the element A with its natural isotopic composition to 1/12 of the mass of an atom of carbon-12.

For example, $A_r(Br)$ = 79.916; $A_r(Zr)$ = 91.22

RELATIVE MOLECULAR MASS

The *relative molecular mass*, M_r (formerly called the 'molecular weight'), is the average mass 'per formula' of the compound with its constituent atoms in their natural isotopic composition to 1/12 of the mass of an atom of carbon-12. For example,

$M_r(KCl)$ = 74.56; $M_r(Na_2H_2Y.2H_2O)$ = 372.23 (for the hydrated sodium salt of EDTA (H_4Y). For 'ferric alum' we have

$M_r((NH_4)_2SO_4.Fe_2(SO_4)_3.24H_2O)$ = 964.42, whereas if the halved formula is adopted,
$M_r((NH_4)Fe(SO_4)_2.12H_2O)$ = 482.21

Note Every *physical quantity* is the product of a *numerical value* (a pure number) and a *unit*. It will be appreciated that when the *molar mass*, $M(X)$, is correctly expressed in its proper units (g mol^{-1}) the pure number is identical with that for the *relative atomic mass*, A_r (formerly atomic weight), or *relative molecular mass*, M_r (formerly molecular weight).

Examples
$M(Ca^{2+})$ = 40.08 g mol^{-1}; $A_r(Ca^{2+})$ = 40.08
$M(Et_2O)$ = 74.124 g mol^{-1}; $M_r(Et_2O)$ = 74.124

USAGE OF TERMS 'EQUIVALENT' AND 'NORMAL'

The analytical chemist will, therefore, find no numerical changes when replacing the older concept of 'molecular weight' by the modern term 'molar mass', but he must never forget that the latter term must be associated with the appropriate unit (g mol^{-1}).

CONCENTRATION

The amount-of-substance concentration (symbol c) is the amount of substance divided by the volume of solution*. The SI base unit is mol m^{-3}, but the practical units are mol dm^{-3} or or mol l^{-1}. These two are in fact identical since the litre has been redefined as being 1 dm^3

Examples

$c(HCl)$ = 0.1 mol l^{-1}
$c(H_3PO_4)$ = 0.5 mol dm^{-3}
$c(S_2O_3^{2-})$ = 3 mmol l^{-1}

Note (i) There is seldom in practice any confusion caused by abbreviating the description 'amount-of-substance concentration' to the more familiar 'concentration'.

(ii) A solution with an amount-of-substance concentration of 0.1 mol dm^{-3} is often called a 0.1 molar solution and written as a 0.1 M solution. See ref. (2), page 6, footnote (5).

(iii) The term molality (amount of substance of X divided by the mass of solvent; unit mol kg^{-1}) will be preferred when quantitative measurements are carried out under non-isothermal conditions, because the molality, but not the concentration, is independent of temperature.

(iv) For linguistic reasons the approved term *molality* could easily be confused with 'molarity', a term formerly used - and still very widely used - to denote concentration (generally in terms of 'gram-molecules per litre'). Since 'molarity' is fully covered by the term *amount-of-substance concentration* it is clearly redundant and it has been recommended that its use should be abandoned. Use of the adjective *molar* is, however, still permitted (cf. Note (ii) above).

To summarise so far: the practising chemist need only realise that many of the physical quantities which he has been accustomed to use have not changed their numerical values, provided they are now associated with particular SI units, and that these changes are concomitant with certain changes in terminology and in the symbols to be used.

Before proceeding further, two other general terms, already defined, should be restated here.

STANDARD SOLUTION A standard solution is one having an accurately known concentration of the active substance, or an accurately known titre[4].

---

* Since volume, V, is a function of temperature, the concentration, c, must also be a function of temperature. Strictly speaking, this should always be specified and (ideally) the operating temperature for a titrimetric analysis should be that for which glassware has been calibrated and at which solutions have been made up. The practising analyst will, of course, be aware of the effect of temperature variations on his results and can make the appropriate corrections if these are justified by the level of accuracy sought for. When a high degree of accuracy and precision is essential, weight burettes may be preferred.

EQUIVALENCE-POINT The point in a titration at which the amount of titrant added is chemically equivalent to the amount of substance titrated[4]. The terms stoicheiometric point and theoretical end-point are synonymous with equivalence-point[4].

EQUIVALENCE

The amount of substance reacting according to equation (1) should clearly be measured in terms of the appropriate unit, the mole, and all concentrations should preferably be expressed in mol dm^{-3} or mol l^{-1}.

Taking a specific case of the general equation (1)

$$HCl + NaOH = NaCl + H_2O \qquad (2)$$

the equivalence point will be reached when each elementary entity of HCl has reacted with just one of NaOH; this will correspond to equal volumes of solutions of HCl and NaOH if they are of equal concentrations, *i.e.*

$$c(HCl) = c(NaOH)$$

For the reaction

$$2NaOH + H_2SO_4 = Na_2SO_4 + 2H_2O \qquad (3)$$

it is clear that each reacting entity of sulphuric acid will be equivalent to two of sodium hydroxide at the equivalence point. If this reaction is rewritten in the form

$$NaOH + \tfrac{1}{2}H_2SO_4 = \tfrac{1}{2}Na_2SO_4 + H_2O \qquad (3a)$$

we see that the amount of the two reactants would be equivalent when $n(NaOH) = n(\tfrac{1}{2}H_2SO_4)$, for the definition of the mole permits us to refer to *any* specified entity, *e.g.* $(\tfrac{1}{2}H_2SO_4)$. Clearly if $n(NaOH) = 1$ mol this amount has a mass of 40 g, and if $n(\tfrac{1}{2}H_2SO_4) = 1$ mol it has a mass of 49 g; these symbols clearly express the quantitative relationship in the form to which we are already accustomed.

The appropriate standard solutions which would neutralise each other when mixed in equal volumes will have $c(NaOH) = c(\tfrac{1}{2}H_2SO_4) =$ (say) 0.1 mol l^{-1} and can be specified as having concentrations of 4.0 g l^{-1} and 4.9 g l^{-1} respectively.

It would appear possible to generalize this approach by writing equation (1) in the form

$$A + (\nu_B/\nu_A)B \longrightarrow \text{products} \qquad (1a)$$
$$\text{where } \nu_A \geqslant \nu_B$$

which signifies that one entity of species A will be equivalent to (ν_B/ν_A) entity of B (in this particular reaction). Let us denote the ratio (ν_B/ν_A) by the symbol $f_{eq}(B)$, and term it the *equivalence factor* of B. The equivalence factor (which will take the form of an integral fraction equal to or less than unity) is a pure number which can be calculated from a knowledge of the stoicheiometry of the given reaction (but see below* and Appendix A). In the above example $f_{eq}(A) = 1$.

---

* Some analytical reactions that proceed reproducibly under carefully controlled conditions cannot be represented by equations in which ν_A and ν_B are small integers and an empirical 'factor' must be used.

USAGE OF TERMS 'EQUIVALENT' AND 'NORMAL'

According to equation (1a) one entity of A will have reacted with $f_{eq}(B)$ entity of B at the equivalence point. However, since the whole concept allows us to refer to any designated species we can equally well refer to one entity of A as reacting with one entity of $f_{eq}(B)$ B.

It is convenient to relate the amount of substance and the amount of substance concentration of an analytical reagent X to the entity $f_{eq}(X)X$ rather than the entity X itself when $f_{eq} \neq 1$. This may be called the equivalent entity of X or the equivalent of X. It is not a physical quantity like $n(f_{eq}(X)X)$ e.g. $n(\frac{1}{2}H_2SO_4)$ but is in the nature of a chemical formula which is a common way of designating an entity. With this choice, if $c(f_{eq}(A)A) = c(f_{eq}(B)B)$ then the equivalence point is reached when equal volumes of the solutions have been mixed. Where $f_{eq}(X)$ varies with the nature of the reaction, this reaction must be specified.

Thus for reaction (3) we have
$$n(NaOH) = n(\tfrac{1}{2}H_2SO_4)$$
at the equivalence point and similarly for the reactions

$$H_3PO_4 + 2KOH = K_2HPO_4 + 2H_2O \qquad (4)$$
$$n(KOH) = n(\tfrac{1}{2}H_3PO_4)$$

whereas for

$$H_3PO_4 + KOH = KH_2PO_4 + H_2O \qquad (5)$$
$$n(KOH) = n(H_3PO_4)$$

and for

$$H_3PO_4 + 3AgOH = Ag_3PO_4(s) + 3H_2O \qquad (6)$$
$$n(AgOH) = n(\tfrac{1}{3}H_3PO_4)$$

As is well known, the equivalent of a substance is not invariable and may change according to the reaction in which it is involved. The reaction must, therefore, always be specified unless there is no possibility of ambiguity in the context.

EQUIVALENCE FOR ACID-BASE REACTIONS

If for any reason the reaction

$$H_2SO_4 + Ca(OH)_2 = CaSO_4 + 2H_2O \qquad (7)$$

were to be conducted titrimetrically to an acid-base end-point we would have $n(H_2SO_4) = n(Ca(OH)_2)$. If one were to use the general definition of equation (1), then $f_{eq}(H_2SO_4)$ would be 1 in contrast to equation (3a) which gives $f_{eq}(H_2SO_4) = \frac{1}{2}$. To avoid such inconsistency it is recommended that one refer all neutralization reactions to a common basis which is covered by the following definition.

The *equivalent* of an acid (or a base) is that entity which, in a specified reaction, would release (or combine with), or be in any other appropriate way equivalent to, 1 entity of titratable hydrogen ions.

On this basis solutions with amounts of substance denoted by $n(HCl)$, $n(NaOH)$, $n(\tfrac{1}{2}H_2SO_4)$, $n(\tfrac{1}{2}H_2C_2O_4)$, $n(H_3PO_4)$ for reaction (5), $n(\tfrac{1}{2}H_3PO_4)$ for reaction (4) and $n(\tfrac{1}{3}H_3PO_4)$ for reaction (6) and where $n(X) = 1$ mol will each contain one mole of equivalent acid or base.

EQUIVALENCE FOR REDOX REACTIONS

Again, since the mole can refer to any specified entity it will be convenient in redox reactions to correlate the amount of reactant with the number of electrons per mole which it combines with or releases.

The *equivalent* of an oxidising (or a reducing) agent is that entity which in a single specified reaction can accept, release, or be in any other way equivalent to one entity of electrons. On this basis solutions with amounts of substance denoted by $n(Fe^{2+})$, $n((NH_4)_2SO_4.FeSO_4.6H_2O)$, $n(\frac{1}{5}KMnO_4)$, $n(\frac{1}{2}H_2C_2O_4)$, $n(\frac{1}{2}Na_2C_2O_4)$, $n(\frac{1}{4}KH_3(C_2O_4)_2.2H_2O)$, $n(\frac{1}{6}K_2Cr_2O_7)$, $n(\frac{1}{2}I_2)$, $n(I^-)$, $n(S_2O_3^{2-})$ etc., where $n(X) = 1$ mol in each example will each contain one mole of equivalent oxidizing or reducing agent.

If one mole of the equivalent of X is dissolved in one litre of solution this particular standard solution can be termed a normal solution which can be defined as follows. A *normal solution* of the species X has an amount-of-substance concentration $c(f_{eq}(X)X) = 1$ mol dm^{-3} (or 1 mol l^{-1}), where $f_{eq}(X)$ is the equivalence factor for X in the reaction under consideration, which must be specified unless there is no ambiguity in the context.

Such a solution can be referred to as 1 normal X (and written N X).

Examples

$c(NaOH) = 1$ mol l^{-1} is the concentration of a normal solution of sodium hydroxide (N NaOH) containing 40 g l^{-1};

$c(\frac{1}{2}H_2SO_4) = 1$ mol l^{-1} is the concentration of a normal solution of sulphuric acid (N H$_2$SO$_4$) containing 49 g l^{-1};

$c(\frac{1}{5}KMnO_4)$ is the concentration of a normal solution of potassium permanganate (for reactions in acid solution) (N KMnO$_4$) containing 31.606 g l^{-1})

$c(\frac{1}{2}I_2) = 1$ mol l^{-1} is the concentration of a normal solution of iodine (N I$_2$) containing 126.92 g l^{-1}.

Note that while all these normal solutions have the same amount-of-substance concentration (1 mol l^{-1}) of the equivalent of the named substance they may differ in content from the corresponding molar solutions because different entities are specified (*e.g.* $\frac{1}{2}$ H$_2$SO$_4$ rather than H$_2$SO$_4$).

Similarly it is possible to designate the content of other standard solutions as, for example, 0.1268 N H$_2$SO$_4$, *i.e.* 0.0634 M H$_2$SO$_4$, etc.

In principle then, there can be no difficulty in specifying the equivalent amount of a substance taking part in a specified reaction and deriving the corresponding amount-of-substance concentration of a normal solution, provided the equivalence factor can be determined. This can present some small difficulties which are discussed in Appendix A, to which is appended a Table giving values of $f_{eq}(X)$ for a number of common analytical reactions of interest in titrimetry.

In complexometric titrations the essential reaction is most commonly the formation of a 1:1 complex between a cation M^{n+} and a ligand Y^{4-} and there is no change in the oxidation state of either reactant, nor are there relevant changes due to the release of protons from the conjugate acid of the ligand which may be used as the titrant. As for example:

$$M^{n+} + H_jY^{(4-j)-} = MY^{(4-n)-} + jH^+ \text{ (taken up in a buffer solution)}.$$

In this case, normal solutions of the reactants have the same concentrations as molar solutions of the same reacting species. There is clearly no advantage here in using the concept of a normal solution.

In other cases, however, it is quite possible to give a rigorous definition for a normal solution of the same substance acting as a titrant in two different reactions because two different values for $f_{eq}(X)$ may be involved. For example (see Appendix), a normal solution of potassium iodate is based on $f_{eq}(KIO_3) = \frac{1}{6}$ for reaction with potassium iodide in dilute acid solution (equation (11)), but $f_{eq}(KIO_3) = \frac{1}{4}$ for the reaction in 2N HCl (Andrews titration); equation (10). If the same standard solution of potassium iodate were to be used for both of these reactions, mistakes could occur if solutions were labelled merely in terms of their normality, unless the value of $f_{eq}(KIO_3)$ is also specified.

APPENDIX A

METHODS OF COMPUTING THE EQUIVALENCE FACTOR, $f_{eq}(X)$

Textbooks of analytical chemistry (written before the widespread adoption of SI) devote a considerable amount of space to discussions of how to arrive at the composition of a normal solution of a titrant for a given reaction. In acid-base titrations the number of replaceable hydrogen atoms has invariably formed the basis for calculating equivalents. In redox reactions the earliest approach was tied to this by way of oxidisable hydrogen or 'available oxygen'. For example, the once familiar $2\ KMnO_4 = K_2O.2MnO.5O$ when $2\ KMnO_4 = 5\ O = 10\ H$ leading to $f_{eq}(KMnO_4) = \frac{1}{5}$. The same result is obtained by the overall stoicheiometry of equation (9). A more recent approach is to base the equivalence factor on the change in oxidation number $|\Delta Z|$ (6). Reference to the following five reactions shows that these two approaches are not always adequate and that they can lead to inconsistent values for $f_{eq}(X)$.

$$2NaOH + H_2SO_4 = Na_2SO_4 + 2H_2O \tag{3}$$

$$10FeSO_4 + 2KMnO_4 + 8H_2SO_4 = K_2SO_4 + 2MnSO_4 + 5Fe_2(SO_4)_3 + 8H_2O \tag{9}$$

$$M^{n+} + H_jY^{(4-j)-} = MY^{(4-n)-} + jH^+ \tag{8}$$

$$KIO_3 + 2KI + 6HCl = 3ICl + 3KCl + 3H_2O \tag{10}$$

$$KIO_3 + 5KI + 6HCl = 3I_2 + 6KCl + 3H_2O \tag{11}$$

Using the relationship $f_{eq}(B) = \nu_B/\nu_A$ consideration of stoicheiometry leads correctly to $f_{eq}(H_2SO_4) = 1/2$, $f_{eq}(KMnO_4) = 2/10$ and $f_{eq}(H_jY) = 1/1$ for equations (3), (9) and (8), but gives the wrong values for the two reactions involving potassium iodate.

Considerations based on changes in oxidation number are of course inapplicable to equations (3) and (8) but give the correct value for (9) when written in the form

$$MnO_4^- + 8H^+ + 5e^- = Mn^{2+} + 4H_2O \tag{9a}$$

since the change in oxidation number[5] is from +7 to +2 whence $|\Delta Z| = 5$ and the equivalence factor is 1/5. For equation (10) (Andrew's titration) we have

$$IO_3^- + 6H^+ + 4e^- = I^+ + 3H_2O \tag{10a}$$

leading (correctly) to $|\Delta Z| = 5 - (+1) = 4$, and $f_{eq}(KIO_3) = 1/4$. However, for equation (11) we can write

$$IO_3^- + 6H^+ + 5e^- = \tfrac{1}{2}I_2 + 3H_2O \text{ with } |\Delta Z| = |5 - 0| = 5, \text{ and}$$

$f_{eq}(KIO_3) = 1/5$ which is incorrect. (It would be appropriate only if the reaction were followed potentiometrically by noting changes in I_2). The complication arises here from the fact that the element iodine is involved in both titrant and analyte and in different oxidation states. Furthermore the actual titrimetric reaction involves further titration of the liberated elementary iodine to iodide with thiosulphate, thus:

$$I_2 + 2Na_2S_2O_3 = 2NaI + Na_2S_4O_6 \qquad (12)$$

The overall stoicheiometry is thus $KIO_3 = 3I_2 = 6NaI$ (or $6e^-$) leading to $f_{eq}(KIO_3) = 1/6$.

In reactions of copper (II) we have

$$Cu^{2+} + 2e^- = Cu \qquad (13)$$

in electrogravimetry leading to $f_{eq}(Cu^{2+}) = 1/2$ for the cupric ion. In contrast $f_{eq}(Cu^{2+}) = 1$ will be appropriate for the reaction.

$$2Cu^{2+} + 4I^- = 2CuI + I_2 \qquad (14)$$

since $2Cu^{2+} = I_2 = 2I^-$ (or $2e^-$), with $f_{eq}(Cu^{2+}) = 2/2 = 1$.

Similarly we note different titrimetric reactions involving Ag^+ which can lead to different values for the equivalent depending upon the actual reaction under consideration. For example:

$$Ag^+ + Cl^- = AgCl \qquad (15)$$
$$Ag^+ + 2CN^- = Ag(CN)_2^- \qquad (16)$$
$$2Ag^+ + Ni(CN)_4^{2-} = 2Ag(CN)_2^- + Ni^{2+} \qquad (17)$$

(Ni^{2+} determined complexometrically with EDTA)

Provided the stoicheiometry of an effective *overall* analytical reaction is considered, the analyst will have no difficulty in computing the appropriate value for the equivalence factor $f_{eq}(X)$. It then becomes readily possible to compute the amount of substance of the equivalent of X, *i.e.* $n(f_{eq}(X) X)$ and the corresponding amount-of-substance concentration $c(f_{eq}(X) X)$ the favoured units being mol and mol dm^{-3} (or mol l^{-1}) respectively.

Where $f_{eq}(X) = 1$ there is clearly no point in preferring the use of normal to molar solutions. Where $f_{eq}(X)$ is less than one the practical advantage of specifying normal solutions derives entirely from the ability to work with comparable volumes in achieving the equivalence-point: against this must be set the disadvantages and possibilities for error when using a reagent (*e.g.* KIO_3 which has different equivalence in different reactions) for which the same standard solution could well be applied.

Where the use of normal solutions is preferred it is recommended that the designation of the standard solution should be explicit on the lines of the following examples.

(a) 0.1257 N sulphuric acid; $f_{eq}(H_2SO_4) = 1/2$; 6.114 g l^{-1} H_2SO_4
(b) 0.1030 N potassium dichromate; $f_{eq}(K_2Cr_2O_7) = 1/6$; 5.0504 g l^{-1} $K_2Cr_2O_7$
(c) 0.1 N potassium iodate; $f_{eq}(KIO_3) = 1/4$; 5.351 g l^{-1} KIO_3;
   reaction $KIO_3 + 2KI + 6HCl \longrightarrow 3ICl + 3KCl + 3H_2O$

Table I gives a selection of titrimetric reactions and the appropriate values for $f_{eq}(X)$.

USAGE OF TERMS 'EQUIVALENT' AND 'NORMAL'

TABLE I

REACTANTS	EQUATION	EQUIVALENTS, $f_{eq}(X)\ X$
HCl , NaOH	HCl + NaOH = NaCl + H$_2$O	1 HCl , 1 NaOH
H$_2$SO$_4$, KOH	H$_2$SO$_4$ + 2 KOH = K$_2$SO$_4$ + 2 H$_2$O	$\frac{1}{2}$ H$_2$SO$_4$, 1 KOH
H$_3$PO$_4$, KOH	H$_3$PO$_4$ + KOH = KH$_2$PO$_4$ + H$_2$O	1 H$_3$PO$_4$, 1 KOH
H$_3$PO$_4$, KOH	H$_3$PO$_4$ + 2 KOH = K$_2$HPO$_4$ + 2 H$_2$O	$\frac{1}{2}$ H$_3$PO$_4$, 1 KOH
H$_3$PO$_4$, KOH	H$_3$PO$_4$ + 3 KOH = K$_3$PO$_4$ + 3 H$_2$O	$\frac{1}{3}$ H$_3$PO$_4$, 1 KOH
H$_2$C$_2$O$_4$, NaOH	H$_2$C$_2$O$_4$ + 2 NaOH = Na$_2$C$_2$O$_4$ + 2 H$_2$O	$\frac{1}{2}$ H$_2$C$_2$O$_4$, 1 NaOH
HCl , Ba(OH)$_2$	2 HCl + Ba(OH)$_2$ = BaCl$_2$ + 2 H$_2$O	1 HCl , $\frac{1}{2}$ Ba(OH)$_2$
R.COOH , Ba(OH)$_2$	2 R.COOH + Ba(OH)$_2$ = Ba(R.COO)$_2$ + 2 H$_2$O	1 R.COOH , $\frac{1}{2}$ Ba(OH)$_2$
HNO$_3$, Na$_2$CO$_3$	2 HNO$_3$ + Na$_2$CO$_3$ = 2 NaNO$_3$ + H$_2$O + CO$_2$	1 HNO$_3$, $\frac{1}{2}$ Na$_2$CO$_3$
HNO$_3$, Na$_2$CO$_3$	HNO$_3$ + Na$_2$CO$_3$ = NaHCO$_3$ + NaNO$_3$	1 HNO$_3$, 1 Na$_2$CO$_3$
AgNO$_3$, KCN	2 KCN + AgNO$_3$ = KAg(CN)$_2$ + KNO$_3$	$\frac{1}{2}$ AgNO$_3$, 1 KCN
AgNO$_3$, KCN	KCN + AgNO$_3$ = AgCN + KNO$_3$	1 KCN , 1 AgNO$_3$
CuSO$_4$, KI	CuSO$_4$ + 2 KI = CuI + $\frac{1}{2}$ I$_2$ + K$_2$SO$_4$	1 CuSO$_4$, 1 KI
KMnO$_4$, FeSO$_4$	2 KMnO$_4$ + 10 FeSO$_4$ + 8 H$_2$SO$_4$ = K$_2$SO$_4$ + 2 MnSO$_4$ + 5 Fe$_2$(SO$_4$)$_3$ + 8 H$_2$O	$\frac{1}{5}$ KMnO$_4$, 1 FeSO$_4$
KMnO$_4$, H$_2$O$_2$	2 KMnO$_4$ + 5 H$_2$O$_2$ + 3 H$_2$SO$_4$ = K$_2$SO$_4$ + 2 MnSO$_4$ + 8 H$_2$O + SO$_2$	$\frac{1}{5}$ KMnO$_4$, $\frac{1}{2}$ H$_2$O$_2$
KMnO$_4$, H$_2$C$_2$O$_4$	2 KMnO$_4$ + 5 H$_2$C$_2$O$_4$ + 3 H$_2$SO$_4$ = K$_2$SO$_4$ + 2 MnSO$_4$ + 10 CO$_2$ + 8 H$_2$O	$\frac{1}{5}$ KMnO$_4$, $\frac{1}{2}$ H$_2$C$_2$O$_4$
KMnO$_4$, As$_2$O$_3$	4 KMnO$_4$ + 5 As$_2$O$_3$ + 12 HCl = 4 KCl + 4 MnCl$_2$ + 5 As$_2$O$_5$ + 6 H$_2$O	$\frac{1}{5}$ KMnO$_4$, $\frac{1}{4}$ As$_2$O$_3$
K$_2$Cr$_2$O$_7$, FeCl$_2$	K$_2$Cr$_2$O$_7$ + 6 FeCl$_2$ + 14 HCl = 2 KCl + 2 CrCl$_3$ + 6 FeCl$_3$ + 7 H$_2$O	$\frac{1}{6}$ K$_2$Cr$_2$O$_7$, 1 FeCl$_2$
Ce(SO$_4$)$_2$, FeSO$_4$	2 Ce(SO$_4$)$_2$ + 2 FeSO$_4$ = Ce$_2$(SO$_4$)$_3$ + Fe$_2$(SO$_4$)$_3$	1 Ce(SO$_4$)$_2$, 1 FeSO$_4$
TiCl$_3$, FeCl$_3$	FeCl$_3$ + TiCl$_3$ = FeCl$_2$ + TiCl$_4$	1 FeCl$_3$, 1 TiCl$_3$
SnCl$_2$, Fe$_2$(SO$_4$)$_3$	SnCl$_2$ + Fe$_2$(SO$_4$)$_3$ + 2 HCl = SnCl$_4$ + 2 FeSO$_4$ + H$_2$SO$_4$	$\frac{1}{2}$ SnCl$_2$, $\frac{1}{2}$ Fe$_2$(SO$_4$)$_3$

USAGE OF TERMS 'EQUIVALENT' AND 'NORMAL'

SUMMARY

1. EQUIVALENCE FACTOR $f_{eq}(X)$

The equivalence factor for a reacting component of a specified titrimetric reaction is a pure number derived from consideration of the overall stocheiometry of the reaction.

For a reaction

$$\nu_A A + \nu_B B \longrightarrow \text{Products}$$

where $\nu_A > \nu_B$ the equivalence factor of reagent A, $f_{eq}(A)$, is taken as unity and for B, $f_{eq}(B)$, it is ν_B/ν_A. A consequence of this definition is that $f_{eq}(X)$ is always unity or less than unity. See p. 179.

N.B. Modifications to this general rule exist for acid-base and oxidation-reduction titrations.

In the case of a reaction that can be clearly identified as acid-base, the equivalence factor for each reacting component must be related to one entity of titratable hydrogen ions.

Thus for a reaction

$$H_2X + B(OH)_2 = BX + 2H_2O$$
$$f_{eq}(H_2X) = 1/2 = f_{eq}(B(OH)_2)$$

In the case of a reaction that can be clearly identified as oxidation-reduction, the equivalence factor for each reacting component must be related to one entity of **transferrable** electrons.

Thus for a reaction

$$M^{n+} + 2e^- = M^{(n-2)+}$$
$$f_{eq}(M^{n+}) = 1/2$$

2. THE EQUIVALENT

The equivalent of a species X is that entity which in a specified reaction would combine with or be in any other appropriate way equivalent in

(a) an acid-base reaction to one entity of titratable hydrogen ions, H^+

or (b) a redox reaction to one entity of electrons, e^-.

In both instances the equivalent can be established from a knowledge of the equivalence factor and the chemical formula of the species and is

$$f_{eq}(X)\ X$$

3. NORMAL SOLUTION

A solution in which the amount-of-substance concentration of the equivalent of the reagent is 1 mol dm^{-3} (*i.e.* 1 mol l^{-1}) may be termed a Normal solution, symbol N.

Decimalised fractions of N may be used, *e.g.* 0.326 N H_2SO_4, that is a solution with $c\ (\tfrac{1}{2}H_2SO_4) = 0.326$ mol l^{-1}.

Bottles containing standard solutions labelled in terms of normality must be labelled clearly and unequivocally to indicate the species and the normality. Because confusion may exist when a reagent has different equivalence factors according to circumstances, the statements of normality must be accompanied by the equivalence factor, *e.g.*

$$0.1 \text{ N KIO}_3 \ ; \ (f_{eq}(\text{KIO}_3) = \tfrac{1}{6})$$
$$0.05 \text{ N KIO}_3 \ ; \ (f_{eq}(\text{KIO}_3) = \tfrac{1}{4})$$

REFERENCES

1. F. Szabadvary, *History of Analytical Chemistry*, 1966; cf. H.M.N.H. Irving, *The Techniques of Analytical Chemistry*, H.M. Stationery Office, London, 1974.
2. *Manual of Symbols and Terminology for Physicochemical Quantities and Units*, IUPAC, 1973 Edition, Butterworths, London, 1975.
3. *IUPAC Information Bulletin No. 36*, August 1974; cf. also *Proceedings of ISO/TC 47 Conference*, Moscow, April 1975.
4. *Pure and Applied Chem.* 1969, 18 (No. 3), 427-436.
5. *Nomenclature of Inorganic Chemistry*, IUPAC 2nd Ed., Butterworths, London, 1971, p.5.
6. *DIN* 32625 (1977).

INDEX

Abbreviations (in thermal analysis), 6.2

Abscissa, positive (in electrochemical sign convention), 20

Absolute temperature, 18.6.1

Absorbance, internal, T.18.5

   peak value of, T.18.5

Absorption factor (in spectrochemistry), 16.5.3, T.18.5

Absorption path-length (in spectrochemistry), T.18.5

Ac polarography, 19.6.4

   higher-harmonic, 19.6.8

      with phase-sensitive rectification, 19.6.9

Accuracy, 1.3.1, 17.2.7

   in spectrochemical work, 17.2.7, 18.4.3.2

   of practical scales of pa_H in 50% methanol and deuterium oxide, 10.8

Acid-base indicator, 8.17.03

Acid-base titration, 8.27.01

Acid form of cation exchanger, 15.2.12

Acidimetric titration, 8.27.02

Acidimetry, 8.01

Acidity constant, 11.3.4.1

Acidity functions, 10.4.2

Active solid, 13.4.04, 14.8.09

   modified, 14.8.10

Activity coefficients,

   mean, 9.5

   molal, 10.4

   of the chloride ion, 9.5

Addition technique, analyte, 18.4.2

Additives, 16.7.7.1, 18.4.1

Additivity of mass spectra, 7.57

Adjusted retention volume, 13.6.03, 14.9.25, 15.2.41

Adsorbate, 4.2.29

Adsorbent, 4.2.28

Adsorption, 4.2.27

Adsorption chromatography, 14.4.1

Adsorption indicator, 8.17.04

* The significance of certain capital letters appearing in the references is as follows: A = Appendix; E = Equation; F = Figure; and T = Table.

INDEX

Aerosol, 18.3.1.1.1

Ageing, 4.2.37

Agglomeration, 4.2.26

Aggregate, 4.2.23

Aggregation, 4.2.24

Air-peak, 13.6.02

Alkalimetric titration, 8.27.03

Alkalimetry, 8.02

a-Parameter (of a spectral line), T.18.5

Alternating-current chronopotentiometry, 19.5.6

Alternating-voltage chronopotentiometry, 19.6.7

Aminopolycarboxylic acid, 8.03

Amperometric end-point detection, 8.12.01

Amperometric titration, 19.4.18
   with two indicator electrodes, 19.4.19

Amperometry, 19.4.15
   differential, 19.4.17
   stirred mercury-pool, 19.4.15
   with two indicator electrodes, 19.4.16

Amphiprotic solvent, measurement of pH in, 10, 10.5

Analysis, spectrochemical, 16, 17, 18

Analysis, titrimetric, 8.30

Analysis element, 16.6.1, 16.6.22

Analyte, 18.4.1

Analyte signal, 18.4.1

Analytical balance, 2.3.01

Analytical calibration, 18.4.2

Analytical calibration functions, 17.3.1

Analytical curve, 16.6.3.2, 17.3.1, 18.4.2

Analytical-curve technique, 18.4.2

Analytical evaluation curves, 17.3.1

Analytical functions, 17.3.1

Analytical result, 18.4.1

Analytical samples, 16.6.1

Angular dispersion, 16.5.2.3

Angular nebulizer, 18.3.1.1.2

Anion effect, 18.4.4.2.2

Anion exchange, 15.2.11

Anion exchanger, 15.2.08
   base form of, 15.2.13

Anion, molecular, 7.43

Anodic-cathodic wave, composite, sign convention for, 20

Anodic current, sign convention for, 20

Anodic stripping, 19.1.4
   controlled-potential coulometry, 19.1.4
   voltammetry, 19.1.4

Anti-Stokes fluorescence, 18.5.1.3

Aperture stop, 16.5.13

INDEX

Apparent concentration, 18.4.4.1

Appearance energy, 7.52

Appearance potential, 7.52

Arc line, 16.7.61

Aspiration, rate of (in flame spectroscopy), 18.3.1.2

Aspirator, 18.3.1.1.1

Assay balance, 2.3.02

Atomic emission spectroscopy, 16

Atomic fluorescence, 18.5.1.3

Atomic line, 18.5.1.3

Atomization 18.3.1.1.1

Atomization efficiency, local (in flame spectroscopy), 18.3.1.2

Atomized fraction, local, 18.3.1.2

Atomizer, 18.3.1.1.1
- carbon (or graphite) cup, 18.3.1.3.3.1
- carbon (or metal) filament, 18.3.1.3.3.1
- metal filament (loop, ribbon or braid), 18.3.1.3.3.1

Auto-ionization, 7.60

Automate, 5.2.14

Automated, 5.2.14

Automatic, 5.2.12

Automatic analysis, 5

Automation, 5.2.15

Automatization, 5.2.13

Automatize, 5.2.13

Auxiliary electrode, 19.19

Av polarography, 19.6.6

Average (mean), 1.4.01, 1.5.05, 1.6.01

Average ligand number, 11.3.5

Back-extract (n.), 12.15

Back-extraction, 12.13

Back-titration, 8.04

Background, 16.6.3.3

Background correction, 17.2.5

Background corrector, 18.3.5

Baker - Sampson - Seidel transformation, 16.8.3.4

Balance, precision of, 2.2.13, 2.3.12

Balance, precision, terminology for, 2.1

Bandwidth, of wavelengths, 16.4.2
- spectral, 16.5.2.2.2, T.18.3.2

Base form of ion-exchanger, 15.2.13

Base peak, 7.53
- intensity relative to, 7.54

Baseline, 13.5.01, 14.7.10
- technique in atomic spectroscopy, 18.4.4.3

Bed volume, 14.9.02, 15.2.20
- capacity, 15.2.23

INDEX

Biamperometric end-point detection, 8.12.02
Biamperometric titration, 19.4.19
Biamperometry, 19.4.16
Bias, 1.4.10
    in atomic spectroscopy, 18.4.3.2
Bifunctional ion-exchanger, 15.2.16
Bipotentiometric titration, 19.4.11
Bipotentiometry, 19.4.9
Black body, 16.4.4
Blackening, 16.8.36
Blank background, 18.4.1
Blank correction (in spectrochemistry), 17.2.5
Blank measure (in flame spectroscopy), 18.4.1
Blank scatter, 18.4.3.1
Blank solution, 18.4.1
Blank titration, 8.05
Blaze angle, 16.5.1.9
Blaze wavelength, 16.5.1.9
Bracketing technique, 18.4.2
Break-through capacity of ion-exchanger bed, 15.2.25
Buffer addition technique, 18.4.4.3
Buffer capacity, 8.06
Buffer index, 8.06
Buffer, ionization (in flame spectroscopy), 18.4.4.3
Buffer, ionic-strength adjustment, 21.1.11
Buffer, spectrochemical, 16.7.7.2, 18.4.4.3
Burner, 18.3.1.1.3
    direct injection, 18.3.1.1.3
    premix type (in flame spectroscopy), 18.3.1.1.3
    reversed direct injection type, 18.3.1.1.3
    slot type, 18.3.1.1.3
    three-slot type, 18.3.1.1.3
Burning velocity (of flames), T.18.3.1
Bypass injector, 13.3.01, 14.8.15

Calibration curve, 21.1.02
Calorimetry, differential scanning, 6.2.11, 6.3.03.6
Capacity, bed volume, 15.2.23
    break-through, 15.2.25
    buffer, 8.06
    of a balance, 2.2.02, 2.3.03
    practical specific, 15.2.24
    theoretical specific, 15.2.21
    volume, 15.2.22
Carbon atomizer, 18.3.1.3.3.1
Carbon-cup atomizer, 18.3.1.3.3.1
Carbon-tube atomizer, 18.3.1.3.3.1
Carrier gas, 13.4.01, 14.8.07, 18.3.2

INDEX

Carrier gas, mean interstitial volocity of, 14.9.21
Cathode-ray polarography, 19.5.16
Cathodic process, sign convention for, 20
Cathodic sputtering, 18.3.1.3.3.2
Cathodic stripping, 19.1.4
Cation exchange, 15.2.10
Cation exchanger, 15.2.07
    acid form of, 15.2.12
Cation, molecular, 7.42
Cation effect, 18.4.4.2.2
Chamber saturation, 14.8.16
Chamber-type nebulizer, 18.3.1.1.2
Characteristic concentration, 18.4.2
Charge-step polarography, 19.5.27
Chelatometric titration, 8.27.04
Chemical equilibrium, 18.6.1
Chemical interference, 18.4.4.2.2
Chemical ionization, 7.30
Chemi-ionization, 7.31
Chemiluminescence, suprathermal type, 18.5.1.1
Chemiluminescent indicator, 8.17.05
Choice of coordinates for electrochemical plots, 20
Chopper, 18.3.3
Chromatogram, 13.5.01, 14.7.01
    differential, F.13.6, F.14.7
    integral, F.13.6, F.14.7
Chromatography (n.), 14.7.04
Chromatograph (v.), 14.7.03
Chromatography, 14.1
    adsorption, 14.4.1
    column, 14.5.1
    displacement, 14.2.3
    elution, 14.2.2
    filament, 14.5.5
    flow-programmed, 14.6.2
    frontal, 14.2.1
    gas-, 13.1, 14.3.1
    gas-liquid, 13.2.02, 14.3.1.1
    gas-solid, 13.2.03, 14.3.1.2
    gel-permeation, 14.4.4
    ion-exchange, 14.4.3
    liquid, 14.3.2
    liquid-gel, 14.3.2.3
    liquid-liquid, 14.3.2.1
    liquid-solid, 14.3.2.2
    nomenclature for, 14
    open-tube, 14.5.2
    paper, 14.5.3

Chromatography (continued)
    partition, 14.4.2
    permeation, 14.4.4
    reversed-phase, 14.6.8
    salting-out, 14.6.3
    temperature-programmed, 14.6.1
    thin-layer, 14.5.4
    two-dimensional, 14.6.7

Chronoamperometry, 19.4.20
    convective, 19.4.22
    double-potential-step, 19.5.25
    dropping electrode, with linear potential (or voltage) sweep, 19.5.16
    polarographic, 19.4.20
    stirred-mercury-pool, 19.4.22
    with linear potential sweep, 19.5.8

Chronocoulometry, 19.4.21
    convective, 19.4.28
    double-potential-step, 19.5.26
    potential-step, 19.4.21

Chronopotentiometric constants, sign convention for, 20

Chronopotentiometric end-point, 8.12.03

Chronopotentiometry, 19.4.12
    alternating-current, 19.5.6
    alternating-voltage, 19.6.7
    current-cessation, 19.5.4
    current-reversal, 19.5.4
    cyclic, 19.5.5
    cyclic current-reversal, 19.5.5
    cyclic current-step, 19.5.5
    derivative, 19.4.13,
    programmed-current, 19.5.3
    with linear current sweep, 19.5.1
    with superimposed alternating current, 19.6.1

α-Cleavage, 7.64

β-Cleavage, 7.65

Coagulation (flocculation), 4.2.25

Coefficient, use of term, 16.2.7
    diffusion, 15.2.35
    partition, 13.6.09, 13.9.01, 14.9.32
    selectivity, 15.2.28

Coherent system of units, 16.2.8

Co-ions, 15.2.06

Collection, 4.2.21

Collector (scavenger), 4.2.22

Collisional half-width (of spectral line), T.18.5

Column, 14.8.01

Column chromatography, 14.5.1

Column, open-tubular, 14.8.04

INDEX

Column, packed, 14.8.03
Column performance, 13.8.01, 14.9.38, 15.2.46
Column temperature, 14.8.24
Column volume, 14.9.01, 15.2.19
Combination electrode, 21.1.1.14
Combined techniques (in thermal analysis), 6.1.05
Comparison solution, 8.07
Compendium, format of, 0.4.1
    scope of, 0.4
Complexes, mixed, 11.3.3
    mononuclear binary, 11.3.2
    polynuclear, 11.3.3, 11.3.5
Compleximetric (complexometric) titration, 8.27.05
Compeximetry (complexometry), 8.08
Complexone (complexan), 8.03
Composite anodic-cathodic wave, sign convention for, 20
Concentration, 16.6.21, 17.2.1, 18.4.1
Concentration distribution ratio, 14.9.33, 15.2.30.1
Concentration index, 16.6.3.2
Concentration ratio, 16.6.22
Concentric nebulizer, 18.3.1.1.2
Concomitants, 16.6.1
Conductimetric end-point, 8.12.04
Conductometric titration, 19.2.2
    high-frequency, 19.2.4
Conductometry, 19.2.1
    Dc, 19.2.1
    high-frequency, 19.2.3
Constant, use of term, 16.2.7
Constituent,
    major, 3.3, 16.6.1
    minor, 3.3
Constituent content, 3.1
Constituent classification, 3.3
Constituents, 16.6.1
    major (in spectrochemical analysis), 16.6.1
Contamination of a precipitate, 4.2.20
    by mechanical entrapment, 4.2.31
    phenomena in precipitation from aqueous solution, 4
Continuum, spectral type, 18.5.1.1
Control titration, 8.09
Controlled flow nebulizer, 18.3.1.1.2
Controlled-current coulometry, 19.4.14
    with potentiometric end-point detection, 19.4.14
Controlled-current potentiometric titration, 19.4.10
Controlled-current potentiometry, 19.4.8
Controlled-potential coulometric titration, 19.4.27
Controlled-potential coulometry, 19.4.27

INDEX

Controlled-potential electrogravimetry, 19.4.29
Controlled-potential electroseparation, 19.4.30
Controlled-potential polarography, 19.12
Convective chronoamperometry, 19.4.22
Convective chronocoulometry, 19.4.28
Conversion factor, 16.7.4
    titrimetric, 8.31
Convolution-integral linear-sweep voltammetry, 19.18
Cooled hollow-cathode lamp, 18.3.1.3.3.2
Cooling curves, T.6.2
Cooling-rate curves, T.6.2
Coordinates, sign conventions in electrochemistry, 20
Coprecipitation, 4.2.34
Corrected retention volume, 13.6.04
Corrected selectivity coefficient, 15.2.29
Correction factor, pressure-gradient, 13.6.05, 14.9.11
Correction of direct weighings, 2.2.15, 2.3.04
Coulometric titration, 8.27.06, 19.4.14
    controlled-potential, 19.4.27
Coulometry, controlled current, 19.4.14
    controlled-potential, 19.4.27
    dropping-electrode, 19.4.23
    polarographic, 19.4.23
Counter electrode, 19.19
Counter-ions, 15.2.03
Crossed electric and magnetic fields, 7.22
Crystalline (ion-selective) electrode, 21.2.1.1
Cumulative formation constant, 11.32
Current, symbols for, 19.10
Current-carrying plasma, 18.3.1.3.2
Current-cessation chronopotentiometry, T. 19.5.4
Current-free plasma, 18.3.1.3.2
Current-reversal chronopotentiometry, T. 19.5.4
    cyclic, T.19.5.5
Current-scanning polarography, T.19.5.2
Current-step chronopotentiometry, T.19.5.4
    cyclic, T.19.5.5
Curve, analytical, 16.6.3.2, 17.3.1, 18.4.2
    calibration, 21.1.02
    ionization efficiency, 7.61
    of growth, 18.5.1.2
Curve corrector, 18.3.5
Curves, in thermal analysis, 6.1.04, 6.2
    cooling, T.6.2.06
    cooling-rate, T.6.2.07
    heating, T.6.2.06, 6.3.03.1
    heating-rate, T.6.2.07, 6.3.03.2
    inverse cooling-rate, T.6.2.08

INDEX

Curves, inverse heating-rate, T.6.2.08, 6.3.03.3

Cyclic chronopotentiometry, T.19.5.5

Cyclic triangular-wave polarography, T.19.5.20

Cyclic triangular-wave voltammetry, T.19.5.21

Cyclotron resonance mass spectrometer, 7.17

Damping constant (of a spectral line), T.18.5

Dark current (of a photodetector), T.18.3.2, 18.3.4

DATA, T.6.2.17

Dc conductometry, T.19.2.1

Dc polarography, T.19.5.12

Dead-stop end-point, 8.12.02

Dead volume, 14.9.10

Deflection, 2.2.07, 2.3.05

    magnetic, 7.08

Degree of dissociation (in atomic spectroscopy), T. 18.6

Degree of ionization (in atomic spectroscopy), T.18.6

Demodulation polarography, T.19.6.10

Density (number), in atomic spectroscopy,

    of free atoms, T.18.6

    of free electrons, T.18.6

    of free ions, T.18.6

    of free molecules, T.18.6

Density (number) of particles in various excitation states, T.18.6

Depression (of a signal), 18.4.4.1

Derivative, 6.1.02, 19.11

Derivative chronopotentiometry, T.19.4.13

Derivative differential thermal analysis, T.6.2.10, 6.3.03.5

Derivative dilatometry, T.6.2.15, 6.3.05.2

Derivative polarography, T.19.5.13

Derivative potentiometric titration, T.19.4.5

    inverse, T.19.4.6

Derivative pulse polarography, T.19.5.23

Derivative thermal analysis, T.6.2.07

Derivative thermogravimetry, T.6.2.05, 6.3.02.4

Derivative voltammetry, 19.5.10

Derivitographic analysis, T.6.2.17

Derivitography, T.6.2.17

Designated volume, 8.10

Desolvated fraction (local), in flame spectroscopy, 18.3.1.2

Desolvation, 18.3.1.1.1

Detection, 14.8.19

Detection limit (in flame spectroscopy), 17.4.1, 18.4.3.2

Detector, 13.3.05

    differential, 13.3.05, 14.8.20

    integral, 13.3.05, 14.8.21

Deuterium oxide, measurement of pH in, 10.8

Deviation, 1.4.02, 1.5.06, 1.6.02

    relative standard, 1.4.06, 1.6.05, 17.2.4

INDEX

Deviation, standard, 1.4.04, 1.5.08, 1.6.03, 17.2.3, 18.4.3.1
Devolatilizer, 16.7.7.4
Dielcometric titration, 19.2.6
Dielcometry, 19.2.5
Dielectrometric titration, 19.2.6
Dielectrometry, 19.2.5
Differential, 6.1.02
Differential amperometry, 19.4.17
Differential chromatogram, F.13.6, F.14.7
Differential detector, 13.3.05, 14.8.20
Differential dilatometry, 6.2.16
Differential polarography, 19.5.14
Differential potentiometric titration, 19.4.4
Differential potentiometry, 19.4.2
Differential pulse polarography, 19.6.3
Differential scanning calorimetry, T.6.2.11, 6.3.03.5
Differential thermal analysis, T.6.2.09, 6.3.03.4
    derivative, T.6.2.10, 6.3.03.5
Differential thermogram, 6.1.04
Differential thermogravimetric analysis, T.6.2.05
Differential thermogravimetry, T.6.2.05
Differential voltammetry, 19.5.11
Diffusion coefficient, 15.2.35
Dilatometry, T.6.2.14, 6.3.05.1
    derivative, T.6.2.15, 6.3.05.2
    differential, T.6.2.16, 6.3.05.2
Diluent, in liquid-liquid extraction, 12.10
Diluent, in spectrochemical analysis, 16.7.7.3
Dilution test, 18.4.3.2
Direct injection burner, 18.3.1.1.3
Direct-line fluorescence, 18.5.1.3
Direct methods (in spectrometry), 18.4.2
Discharge polarography, 19.5.27
Dispersion, linear, 16.5.2.3
Dispersion (of a material), 16.5.2.3
Displacement chromatography, 14.2.3
Dissociation (of flame species), 18.6.1
Dissociation constant (in atomic spectroscopy), T.18.6
Dissociation energy (in spectrochemistry), T.18.6
Dissociation interference (in flame spectroscopy), 18.4.4.2.2
Dissociation potential (in spectrochemistry), T.18.6
Dissolution, 4.2.05
Distribution coefficient, 12.06, 15.2.30.2
    logarithmic, 4.2.33
    volume, 15.2.30.3
Distribution coefficients, 12.06, 14.9.34, 15.2.30.2, 15.2.30.3
Distribution constant, 14.9.32
    overall, 11.5.2

Distribution, laws of, 4.2.33
Distribution ratio, concentration, 11.5.3, 12.06, 14.9.33, 15.2.30.1
    mass, 14.9.35
Doerner and Hoskins equation, 4.2.33
Doppler half-width, of spectral line, T.18.5
Double-beam system, 18.3.3
Double-focusing mass spectrometer, 7.06
Double-potential-step chronoamperometry, 19.5.25
Double-potential-step chronocoulometry, 19.5.26
Double-tone polarography, 19.6.13
Drift, 21.1.04
Drop generator nebulizer, 18.3.1.1.2
Dropping electrode coulometry, 19.4.23
Dropping mercury electrode, 19.10
Dry aerosol, 18.3.1.1.1
Dynamic differential calorimetry, 6.2.09
Dynamic-field mass spectrometry, 7.20
Dynamic thermogravimetric analysis, 6.2.04

Effect, spectroscopic, 18.4.4.2.2
Effective theoretical plate number, 14.9.39
Efficiency of atomization, (local), 18.3.1.2
Efficiency of fluorescence,
    power, T.18.5
    quantum, 18.5.1.3, T.18.5
    total quantum, T.18.5
Efficiency of nebulization, 18.3.1.2
Efficiency of transmission, of optical filters, T.18.3.2
Effluent gas analysis, 6.2.13
Effluent gas detection, 6.2.12
Einstein coefficient for spontaneous emission, T.18.5
Einstein transmission probability, T.18.5
Electrical current (lamp current), T.18.3.2
Electrical flame-like plasmas, 18.3.1.3.2
Electrical measuring systems, 18.3.5
Electroanalytical techniques, 19
Electrode, combination, 21.1.14
    reference, 21.1.09
Electrode memory, 21.1.05
Electrodeless plasma, 18.3.1.3.2
Electrodes, ion-selective, 21, 21.2.5, 21.1.07
    crystalline, 21.2.1.1
    enzyme substrate, 21.2.3.2
    gas-sensing, 21.2.3.2
    heterogeneous membrane, 21.2.1.1.2
    homogeneous membrane, 21.2.1.1.2
    internal reference, 21.1.10
    mobile carrier, 21.2.2.2
        negatively charged, 21.2.2.2

INDEX

Electrodes, ion-selective (continued)
    mobile carrier (continued)
        positively charged, 21.2.2.1
        uncharged, 21.2.2.3
    non-crystalline, 21.2.1.2
    primary, 21.2.1
    rigid matrix, 21.2.1.3
    sensitized, 21.2.3
Electrography, 19.4.26
Electrogravimetry, 19.4.24
    controlled potential, 19.4.29
    internal, 19.4.24
    spontaneous, 19.4.24
Electrometer, vibrating reed, 7.37
Electron energy, 7.26
Electron impact ionization, 7.25
Electroseparation, 19.4.25
    controlled potential, 19.4.30
Eluate, 14.8.18
Eluent, 14.8.06
Elute, 14.7.05
Elution band, 14.7.12
Elution chromatography, 14.2.2
Elution curve, 14.7
Elution, gradient, 14.6.6
    selective, 14.6.4
    stepwise, 14.65
Elution volume, peak, 14.9.24
Emission, spectroscopic, 18.5.1.1
Emissivity, radiant, T.16.4
Emulsion calibration curve, 16.8.3.2
Emulsion calibration function, 16.8.3.2
Energy density, radiant T.16.4, T.18.5
End-point, 8.11
    amperometric, 8.12.01
    biamperometric, 8.12.02, 19.4.19
    chronopotentiometric, 8.12.03
    conductometric, 8.12.04
    dead-stop, 8.12.02
    enthalpimetric, 8.12.11
    fluorimetric, 8.12.05
    high-frequency, 8.12.06
    nephelometric, 8.12.07
    photometric, 8.12.08
    potentiometric, 8.12.09
    radiometric, 8.12.10
    theoretical, 8.13
    thermometric, 8.12.11

End point (continued)
    turbidimetric, 8.12.12
    visual, 8.12.13

Energy density, radiant, 16.4, T.16.4, T.18.5

Energy, ionization, 16.7.4

Energy of excitation, 16.7.4, T.18.6

Enhancement, spectroscopic, 18.4.4.1

Entrapment, mechanical, 4.2.31

Enrichment factor, 12.08

Enthalpimetric endpoint, 8.12.11

Entrance slit height, T.18.3.2

Entrance slit width, T.18.3.2

Enzme substrate electrode, 21.2.3.2

Equilibrium, layer, 14.8.17

Equilibrium constant, 11.2, 11.5.2

Equivalence-point, 8.13

Equivalent, chemical, *Appendix*

Error, percentage, 1.4.09
    titration, 8.28

Errors, 1.4.08, 13.9.02

Evolved gas analysis, 6.2.13, 6.3.04.2

Evolved gas detection, 6.2.12, 6.3.04.1

Excitation, spectroscopic, 18.6.1

Excitation energy, 16.7.4, T.18.6

Excitation interferences, 18.4.4.2.2

Excitation potential, 16.7.4, T.18.6

Experimental considerations, 13.9.03

Exposure, radiant, 16.6.3, 16.8.2.2, T.16.4

Exposure time, 16.7.6.4

Expression of results, 1.4

External standard, 8.44

Extract (n.), 12.12

Extract (v.), 14.7.06

Extractability, 12.07

Extractant, 12.09

Extracting agent, 12.11

Extraction, liquid-liquid, 12.1

Extraction coefficient, 12.06

Extraction constant, 12.3

Extraction indicator, 17.06

Extraction, solvent, 12.1

Factor, enrichment, 12.08
    recovery, 12.07
    separation, 4.2.33, 14.9.36, 15.2.31

Factor weight, 8.14

Faradaic rectification, high-level, 19.6.14

Faraday cup (or cylinder) collector, 7.35

Feed-back system, 5.2.11

Field analyser, magnetic, 7.12, 7.13, 7.14

INDEX

Field analyser, radial electrostatic, 7.09
Field ionization, 7.27
Filament chromatography, 14.5.5
Filter, optical, 18.3.3
Fixed ions, 15.2.04
Flame, 18.1
Flame background, 18.4.1
Flame geometry interferences, 18.4.4.2.2
Flame spectroscopy, 16
    interference in, 18.4.4.1
    ionization in, 18.6.1
    multipass system, 18.3.3
Flame temperature, T.18.3.1
Flames, fuel-rich, 18.3.1.1.3
Flash-back (of flames), 18.3.1.1.3
Flocculation (coagulation), 4.2.25
Flow-programmed chromatography, 14.6.2
Flow-rate, of gases, T.18.3.1
    of unburned gases, T.18.3.1
    volumetric, 14.9.18
Fluorescence, atomic, 18.5.1.3
    anti-Stokes, 18.5.1.3
    direct line, 18.5.1.3
    resonance type, 18.5.1.3
    stepwise line, 18.5.1.3
    Stokes type, 18.5.1.3
Fluorescence efficiency,
    power, T.18.5
    quantum, 18.5.1.3
    total quantum, T.18.5
Fluorescent indicator, 18.17.07
Fluorimetric endpoint, 18.12.05
Flux, 16.A.1
Flux through a monochromator, 16.A.6
Focal length, 16.5.14, 16.5.16
Fog (of photographic plate), 16.6.3.3
Formality, 8.15
Format, of compendium, 0.4.1
Formation constant. cumulative, 11.3.2
    overall, 11.3.1
    stepwise, 11.3.2
Fraction atomized (local), 18.3.1.2
Fraction desolvated (local), 18.3.1.2
Fraction volatilized (local), 18.3.1.2
Fraction, interstitial, 14.9.04
    stationary-phase, 14.9.06
Fragment ion, 7.47
Frequency of modulation, T.18.3.2

INDEX

Frequency, spectral, 16.4.3
Front, solvent, 14.8.22
Frontal chromatography, 14.2.1
Fronting, 14.7.14
Fuel, 18.3.1.1.3
Fuel-rich flames, 18.3.1.1.3

Gas, carrier, 13.4.01, 14.8.07
Gas-chromatography, 13, 13.1, 13.2.01, 14.3.1
    table of terms, 13.10
Gas hold-up volume, 13.6.02, 14.9.09
Gas-liquid chromatography, 13.2.02, 14.3.1.1
Gas-permeable membrane, 21.3.1
Gas-solid chromatography, 13.2.03, 14.3.1.2
Gel-permeation chromatography, 14.4.4
Geometry, Mattauch-Herzog, 7.11
    Nier-Johnson, 7.10
Gerlach's homologous lines, 16.6.3.1
Glossary of terms for precision balances, 2.3
Gradient elution, 14.6.6
Gradient layer (or packing), 14.8.13
Graphite-cup atomizer, 18.3.1.3.3.1
Graphite-rod furnace, 18.3.1.3.3.1
Graphite-tube furnace, 18.3.1.3.3.1
Gravity-fed nebulizer, 18.3.1.1.2
Growth curve, 18.5.1.2

H and D curve, 16.8.3.6
Half-intensity width, T.18.5
    of absorption line, 18.3.2
    of emission line, T.18.3.2
    (collisional), T.18.5
    (Doppler), T.18.5
Heating curves, 6.2.06, 6.3.03.1
Heating-rate curves, 6.2.07, 6.3.03.2
Height equivalent to a theoretical plate, 14.9.40
Height equivalent to an effective theoretical plate, 14.9.41
Height of observation, T.18.3.2
Heterochromatic photometry, 16.8.4.4
Heterogeneous membrane electrode, 21.2.1.1.2
High-frequency conductometric titration, 19.2.4
High-frequency conductometry, 19.2.3
High-frequency end-point, 8.12.06
High-level faradaic rectification, 19.6.14
High-pressure Xenon lamp, 18.3.2
Higher-harmonic ac polarography, 19.6.8
    with phase-sensitive rectification, 19.6.9
Hold-up volume, 13.6.02, 14.9.08
Hollow-cathode discharge, 18.3.1.3.3.2

INDEX

Hollow-cathode lamp, 18.3.2
Homogeneous distribution coefficient, 4.2.33
Homogeneous membrane electrode, 21.2.1.1.1
Homologous line, 16.6.3.1
Hot hollow-cathode, 18.3.1.3.3.2
Hunter and Driffield curve, 16.8.3.6
Hydrodynamic voltammetry, 19.5.9
Hysteresis, 21.1.05

Ilkovič equation, sign convention for, 20
Incremental-charge polarography, 19.5.27
Index, use of term, 16.2.7
Index, buffer, 8.06
    retention, 14.9.42
Indicator, acid-base, 8.17.03
    adsorption, 8.17.04
    chemiluminescent, 8.17.05
    extraction, 8.17.06
    fluorescent, 8.17.07
    metallochromic, 8.17.08
    metallofluorescent, 8.17.09
    mixed, 8.17.10
    one-colour, 8.17.01
    oxidation-reduction (redox), 8.17.11
    precipitation, 8.17.12
    radioactive, 8.17.13
    screened, 8.17.13
    two-colour, 8.17.02
    visual, 8.16
Indicator blank, 8.18
Indicator correction, 8.18
Indicator electrode, 19.16
Indicators (visual), types of, 8.17
Indirect titration, 8.27.07
Inductively coupled plasmas, 18.3.1.3.2
Initial and final temperatures, 14.8.26
Injection temperature, 14.8.25
Injector, bypass, 13.3.01, 14.8.15
    sample, 13.3.01, 14.8.14
Inner zone, 18.3.1.1.3
Instrument (n.), 5.2.06
Instrument (v.), 5.2.09
Instrument indication, 2.2.05, 2.3.06
Instrumental, 5.2.08
Instrumentation, 5.2.07
Integral absorption, 16.2.8, T.18.5
Integral chromatogram, 13.6, 14.7
Integral detector, 13.3.05, 14.8.21
Integrating sphere, 16.A.5.5

INDEX

Integration of flux, 16.A.22
Intensity, 16.6.3
Intensity of radiation, 16.3.1, T.18.5
Intensity of spectral line, T.18.5
Intensity bridge ratio, 16.8.4.4
Intensity calibrating device, 16.8.4
Interconal zone, 18.3.1.1.3
Interface, 4.2.01
Interfacial tension, measurement of, 19.3.1
Interference, in flame spectroscopy, 18.4.4.1
    in mass spectrometry, 7.58
Interference, excitation, 18.4.4.2.2
    flame geometry, 18.4.4.2.2
    ionization, 18.4.4.2.2
    lateral diffusion, 18.4.4.2.2
    mutual, 18.4.4.1
    non-spectral, 18.4.4.2.2
    physical, 18.4.4.2.2
    solute volatilization, 18.4.4.2.2
    spatial distribution, 18.4.4.2.2
    specific, 18.4.4.2.2
    spectral, 18.4.4.2.1
    transport, 18.4.4.2.2
    vapour phase size, 18.4.4.2.2
Interference curve, 18.4.4.1
Interferent (in flame spectroscopy), 18.4.4.1
Interfering lines, 16.6.3.4 , 18.4.4.2.1
Interfering substance, 21.1.08
Internal absorbance, T.18.5
Internal electrogravimetry, 19.4.24
Internal reference electrode, 21.1.10
Internal reference line, 16.6.3
Internal standard, 16.8.4.4 , 14.7.17
Interpretation of measured pH, 9.7, 10.3
Interstitial fraction, 14.9.04
Interstitial velocity, 14.9.20
    mean, of carrier gas, 14.9.21
Interstitial volume, 13.3.04, 14.9.03
Interval, transition, 8.32
Interzonal region, 18.3.1.1.3
Inverse cooling-rate curves, 6.2.08
Inverse derivative potentiometric titration, 19.4.6
Inverse heating-rate curves, 6.2.08, 6.3.03.3
Iodimetric titration, 8.27.08
Iodimetry, 8.27.08
Iodometry, 8.27.08
Ion,
    fragment, 7.47

INDEX

Ion (continued)
    isotopic, 7.48
    molecular, 7.41
    parent, 7.45
    precursor, 7.46
    progenitor, 7.46
    rearrangement, 7.49
Ion-exchange, 15.2.02
    nomenclature for, 15
Ion-exchange chromatography, 14.4.3
Ion-exchange isotherm, 15.2.32
Ion-exchange membrane, 15.2.36
Ion-exchanger, 15.2.01
    bifunctional, 15.2.16
    macroporous, 15.2.18
    monofunctional, 15.2.15
    polyfunctional, 15.2.17
    redox, 15.2.39
    salt form of, 15.2.14
Ion-molecule reaction, 7.59
Ion-selective electrodes, 21, 21.2.5, 21.1.07
    recommendations for, 21.1.07
    classification of, 21.2
Ion-specific electrode, 21.1.07
Ionic activity coefficients, 9.5
Ionic line, 16.7.6.1, 18.5.1.1
Ionic medium, 11.2.1
Ionic strength, 21.3.2
    adjustment buffer, 21.1.11
Ionization (in flame spectroscopy), 18.6.1
Ionization buffer, 18.4.4.3
Ionization by sputtering, 7.33
Ionization, chemical, 7.30
Ionization constant, T.18.6
Ionization efficiency curve, 7.61
Ionization, electron impact, 7.25
Ionization energy, 16.7.4, T.18.6
Ionization, field, 7.27
Ionization interference, 18.4.4.2.2
Ionization, laser beam, 7.34
Ionization potential, 16.7.4, T.18.6
Ionization, spark source, 7.32
    thermal, 7.29
Ionizing voltage, 7.26
Ionogenic groups, 15.2.05
Irradiance, T.16.4
Isobaric weight-change determination, 6.2.02 , 6.3.02.1
Isopotential point , 21.1.18

INDEX

Isotherm, ion-exchange, 15.2.32
    sorption, 15.2.34
Isothermal thermogravimetric analysis, 6.2.03
Isothermal weight-change determination, 6.2.03, 6.3.02.2
Isotopic ion, 7.48

Kalousek polarography, 19.5.24

Laminar flames, 18.3.1.1.3
Lamp current, T.18.3.2
Laser-beam ionization, 7.34
Lateral diffusion interferences, 18.4.4.2.2
Laws of distribution, 4.2.33
Layer equilibration, 14.8.17
Length of total effective prism base, 16.5.1.9
Level of titration, 8.20
Ligand numbers, average, 11.3.5
Light modulation, 18.3.3
    frequency of, T.18.3.2
Limit of detection, in flame spectroscopy, 17.4.1, 18.4.3.2
    with ion-selective electrodes, 21.1.03
Line, atomic, 16.7.6.1, 18.5.13
Line-broadening parameter, T.18.5
Line, spectral, 16.7.6.1, 18.5.1.1
Line wavelength, 16.2.8, T.18.5
Line-width, 16.5.2.2
Linear dispersion, 16.5.2.3
Linear flow, nominal, 14.9.19
Linear sweep voltammetry, 19.5.8
Liquid chromatography, 14.3.2
Liquid extraction, 12.01
Liquid-gel chromatography, 14.3.2.3
Liquid-liquid chromatography, 14.3.2.1
Liquid-liquid distribution, 12, 12.01
Liquid-liquid distribution equilibria, 11.5
Liquid phase, 13.4.02
Liquid-solid chromatography, 14.3.2.2
Liquid volume, 13.3.03
Load, 2.2.01, 2.3.07
Local analysis, 18.3.1.3.3.3
Local efficiency of atomization, 18.3.1.2
Local fraction atomized, 18.3.1.2
Local fraction desolvated, 18.3.1.2
Local fraction volatilized, 18.3.1.2
Logarithmic distribution coefficient, 4.2.33
Lomakin - Scheibe equation, 16.6.3.2
Long tube device, 18.3.1.3.1

INDEX

Low-pressure discharge lamp, 18.3.2
Luminous efficacy, 16.4.5 (b)
Luminous efficiency, 16.4.5 (b)
Luminous flux, 16.4.5 (b)
Luminous quantity, 16.4.5 (a)

McLafferty rearrangement, 7.66
Machine, 5.2.02
Macro, 3.2
Macro component, 4.2.12
Macro sample, 3.2
Macroporous ion-exchanger, 15.2.18
Magnetic deflection, 7.08
Magnetic field analyser, π radian, 7.12
    $\pi/2$ radian, 7.13
    $\pi/3$ radian, 7.14
Major constituent, 3.3, 16.6.1
Marker, 14.7.18
Masking agent, 8.21
Mass analyser, quadrupole, 7.15
Mass distribution ratio, 14.9.35
Mass spectra, additivity of, 7.57
Mass spectrograph, 7.02
    double focusing, 7.07
Mass spectrometer, 7.01
    cyclotron resonance, 7.17
    double focusing, 7.06
    dynamic field, 7.20
    linear accelerator type, 7.18
    prolate trochoidal, 7.21
    single-focusing, 7.05
    static field, 7.19
    time of flight, 7.16
Mass spectrometry, 7.04
    nomenclature of, 7
Mass spectroscope, 7.03
Mass spectrum, 7.39
    background, 7.63
    negative-ion, 7.40
Matrix, 16.6.1
Matrix effect, 16.6.1, 18.4.4.2.2
Matrix, resin, 15.2.09
Mattauch - Herzog geometry, 7.11
Maximum transmission, wavelength of, T.18.3.2
Mean activity coefficient, 9.5
Mean (average), 1.4.01, 1.5.05, 1.6.01, 17.4.1
Mean interstitial velocity of the carrier gas, 14.9.21
Mean molal activity coefficient of sodium chloride, 10.7

INDEX

Meaning of qualifying signs (in gas chromatography), 13.6.10
Measure (in flame spectroscopy), 18.4.1
Measured value, 1.2.1
Mechanical, 5.2.03
Mechanical entrapment, 4.2.31
Mechanism, 5.2.01
Mechanization, 5.2.04
Mechanize, 5.2.05
Medium, 11.2.1
Medium effects, 10.4
Medium, ionic, 11.2.1
Membrane, 21.1.03
    gas-permeable, 21.2.3 1
    ion-exchange, 15.2.36
Mercury-pool electrode, 19.16
Mercury electrode, dropping, 19.10
Meso, 3.2
Meso sample, 3.2
Meso/trace analysis 3.3.04
Metallochromic indicator, 8.17.08
Metallofluorescent indicator, 8.17.09
Metastable decomposition, 7.50
Metastable ion-peak, 7.51
Micro, 3.2
Microanalysis, 3.3
Microchemical analysis, scales of working in, 3
Microchemical balance, 2.3.08
Micro-component, 4.2.13
Microcoulometry, 19.4.23
Microphotometer, 16.8.2.3
Microtrace, 3.3
Micro/trace analysis, 3.3.03
Migration-distance, solvent, 14.8.23
Millicoulometry, 19.4.23
Milligram equivalent of readability, 2.2.12, 2.3.09
Minimal line-width, 16.5.2.2
Minor constituent, 3.3
Mist (in flame spectroscopy), 18.3.1.1.1
Mixed complexes, 11.3.3
Mixed crystals (solid solutions), 4.2.32
Mixed indicators, 8.17.10
Mixed solvents, measurement of pH in, 10
Mobile carrier electrode, 21.2.2
Mobile phase, 13.4.01, 14.8.05, 15.2.48
Modified active solid, 14.8.10
Modified Nernst equation, 21.3.1
Modulation frequency, T. 18.3.2
Modulation polarography, 19.6.12

INDEX

Molal activity coefficients, 10.4
Molecular anion, 7.43
Molecular cation, 7.42
Molecular ion, 7.41
    rearranged, 7.44
Monochromator, 16.5.1.1, 16.A.5.6.1, 18.3.3
Monofunctional ion-exchanger, 15.2.15
Mononuclear binary complexes, 11.3.2
Multiple techniques in thermal analysis, 6.1.05, 6.2.17, 6.3.06
Multipass system (in flame spectroscopy), 18.3.3
Multisweep polarography, 19.5.17
Mutual interferences, 18.4.4.1

Nanotrace, 3.3
Nebulization efficiency, 18.3.1.2
Nebulize, 18.3.1.1.1
Nebulizer, 18.3.1.1.2
    angular, 18.3.1.1.2
    chamber type, 18.3.1.1.2
    concentric, 18.3.1.1.2
    controlled flow, 18,3,1,1,2
    drop-generator type, 18.3.1.1.2
    gravity-fed, 18.3.1.1.2
    pneumatic, 18.3.1.1.1, 18.3.1.1.2
    reflux type, 18.3.1.1.2
    suction type, 18.3.1.1.2
    twin type, 18.3.1.1.2
    ultrasonic type, 18.3.1.1.2
Negative-ion mass spectrum, 7.40
Nephelometric end-point detection, 8.12.07
Nernst equation, modified, 21.3.1
Nernstian response, 21.1.12
Net measure (in flame spectroscopy), 18.4.1
Net retention volume, 13.6.06, 14.9.26
Nier - Johnson geometry, 7.10
No-load indication, 2.2.06, 2.3.10
Nomenclature,
    contrasting concepts of, 0.3.4
    duplication of terms in, 0.3.1
    faulty terms in, 0.2.2
    general principles of, 0.2
    internationality in, 0.3.6
    multiplicity of meanings in, 0.3.2
    new technical words in, 0.2.3.5
    (new) terms in, 0.2.3
    (new) words in, 0.2.3.3
    physical quantities in, 0.3
    (physical) symbols for , 0.3
    related concepts in, 0.3.4

Nomenclature (continued)
    self-explanatory terms in, O.2.1
    simplicity in, O.3.5
    sources of material for, O.1
    use of discoverer's name in, O.2.3.4

Nominal linear flow, 14.9.19

Non-aqueous titration, 8.27.09

Non-crystalline (ion-selective) electrode, 21.2.1.2

Non-faradaic admittance, measurement of, 19.3.2

Non-spectral interferences, 18.4.4.2.2

Normal eye (standard observer), 16.4.4 (b)

Normal solution, *Appendix*

Nucleation, 4.2.15
    rate of, 4.2.16

Nucleus, 4.2.14

Null-current potentiometric titration, 19.4.3

Null-current potentiometry, 19.4.1

Number of rules (of diffraction grating), 16.5.1.9

Number in series, 1.5.03

Number of variates, 1.4.07

Numerical aperture, 16.A.5.2

Observation height (in flame spectroscopy), T.18.3.2

Occlusion, molecular, 4.2.30

One-colour indicator, 8.17.01

Open-tube chromatography, 14.5.2

Open-tubular column, 14.8.04

Operator, 9.2

Operational pH scale, 9.3, 10.2
    for amphiprotic solvents, 10.6

Optical conductance, 16.5.3.2, 16.A.1, T.18.3.2

Optical density, 16.8.3.6

Optical filters, 18.3.3

Order of spectrum, 16.5.1.9

Ordinate, positive (in electrochemistry), 20

Organic effect, 18.4.4.2.2

Oscillator strength, T.18.5

Oscillographic polarography, 19.5.7
    single-sweep, 19.5.16

Oscillography, 19.5.7

Ostwald ripening, 4.2.38

Outer zone (in flames), 18.3.1.1.3

Overall distribution constants, 11.5.2

Overall formation constants, 11.3.1

Oxidant (in flames), 18.3.1.1.3

Oxidation-reduction indicator, 8.17.11

Oxidation-reduction titration, 8.27.10

pa_H, 9.8

Packed column, 14.8.03

INDEX

Packing, 14.8.02

Paper chromatography, 14.5.3

Parameter, photographic, 16.8.3.1

Parent ion, 7.45

Partition chromatography, 14.4.2

Partition coefficient, 13.6.09, 13.9.01, 14.9.32

Partition constant, 12.05

Partition function, T.18.6

Parts per billion, 16.6.2.1

Parts per million, 16.6.2.1

Path length, absorption, T.18.5

Peak absorbance, T.18.5

Peak, 13.5.02, 13.6.02, 14.7.11, 15.2.42

Peak-area, 13.5.02, 14.9.13, 15.2.44

Peak base, 13.5.02, 14.9.12, 14.2.43

Peak elution volume, 14.9.24

Peak height in chromatography, 13.5.02, 14.9.15

Peak height (in mass spectrometry), 7.55

Peak-maximum, 14.9.14

Peak resolution, 13.8.02, 14.9.37, 15.2.47

Peak width, 13.5.02, 14.9.16, 15.2.45
    at half-height, 13.5.02, 14.9.17

Percentage error, 1.4.09

Percentage recovery, 1.4.11

Period, spectral, 16.4.3

Permeation chromatography, 14.4.4

Permselectivity, 15.2.37

pH, approximate interpretation of, 9.5, 10.3, 10.6
    definition of, 9.3, 10.2
    interpretation of, in aqueous solutions, 9.3, 9.7
    measurements in amphiprotic and mixed solvents, 10
    medium effects, 10.4

pH of standard solutions, 9.4, 9.6

pH, standards for measuring, 9.4, 10.8

pH unit for amphiprotic solvents, 10.2

Phase, liquid, 13.4.02
    mobile, 13.4.01, 14.8.05, 15.2.48
    stationary, 13.4.05, 14.8.08
    volume of, 14.9.05

Phase-ratio, 14.9.07

Phase titration, 8.27.11

Photocurrent, 18.3.4, T.18.3.2

Photodetector, dark current of, T.18.3.2, 18.3.4

Photodetectors, 18.3.4

Photographic parameter, 16.8.3.1

Photographic plate recording, 7.38

Photoionization, 7.28

Photometric end-point, 8.12.08

Photometry, 16.4.5
    heterochromatic, 16.8.4.4
    quasi-monochromatic, 16.8.4.2
Physical equilibrium (in flames), 18.6.1
Physical interferences (in flame spectroscopy), 18.4.4.2.2
Physical quantity, 16.2.2
π radian magnetic field analyser, 7.12
$\pi/2$ radian magnetic field analyser, 7.13
$\pi/3$ radian magnetic field analyser, 7.14
Picotrace, 3.3
Plasma, current-carrying, 18.3.1.3.2
    current-free, 18.3.1.3.2
    electrical flame-like, 18.3.1.3.2
    electrodeless, 18.3.1.3.2
    inductively coupled, 18.3.1.3.2
    single electrode, 18.3.1.3.2
Plasma jet, 18.3.1.3.2
Plate number, theoretical, 13.8.01, 14.9.38
Plate, support, 14.8.12
Pneumatic nebulizer, 18.3.1.1.1, 18.3.1.1.2
Polarographic chronoamperometry, 19.4.20
Polarographic coulometry, 19.4.23
Polarographic diffucion current constants, 20
Polarographic titration, 19.4.18
Polarography, 19.5.12
    ac, 19.6.4
        higher-harmonic, 19.6.8
            with phase-sensitive rectification, 19.6.9
    av, 19.6.6
    cathode-ray, 19.5.16
    charge-step, 19.5.27
    current-scanning, T.19.5.2
    cyclic triangular-wave, 19.5.20
    dc, 19.5.12
    demodulation, 19.6.10
    derivative, 19.5.13
    derivative pulse, 19.5.23
    differential, 19.5.14
    differential pulse, 19.6.3
    discharge, 19.5.27
    double-tone, 19.6.13
    higher-harmonic ac, 19.6.8
        with phase-sensitive rectification, 19.6.9
    incremental charge, 19.5.27
    Kalousek, 19.5.24
    modulation, 19.6.12
    multisweep, 19.5.17
    oscillographic, 19.5.7

INDEX

Polarography (continued)
    pulse, 19.5.22
        derivative, 19.5.23
        differential, 19.6.3
    radiofrequency, 19.6.11
    semi-integral, 19.1.8
    single sweep, 19.5.16
        oscillographic, 19.5.16
    square-wave, 19.6.5
    staircase, 19.6.2
    Tast, 19.5.15
    triangular-wave, 19.5.18
        cyclic, 19.5.20
Polarometric titration, 19.4.18
Polychromator, 16.5.1.1, 18.3.3
Polyfunctional ion-exchanger, 15.2.17
Polymers, redox, 15.2.38
Polynuclear complexes, 11.3.3, 11.3.5
Positive abscissa (in electrochemistry), 20
Positive ordinate (in electrochemistry), 20
Postprecipitation, 4.2.35
Potential of excitation, 16.7.4, T.18.6
Potential of ionization, 16.7.3, T.18.6
Potential-step chronocoulometry, 19.4.21
    double, 19.5.26
Potentiometric end-point, 8.12.09
Potentiometric selectivity coefficient, 21.1.15
Potentiometric titration, 19.4.3
    controlled-current, 19.4.10
        with two indicator electrodes, 19.4.11
    derivative, 19.4.5
    differential, 19.4.4
    inverse derivative, 19.4.6
    null-current, 19.4.3
    second-derivative, 19.4.7
    zero-current, 19.4.3
Potentiometry, 19.4.1
    controlled-current, 19.4.8
        with two indicator electrodes, 19.4.9
    differential, 19.4.2
    null-current, 19.4.1
    precision null-point, 19.4.2
    zero-current, 19.4.1
Power efficiency of fluorescence, T.18.5
Practical measurement of pH in amphiprotic and mixed solvents, 10
Practical specific capacity, 15.2.24
Pre-arc (pre-spark) period, 16.7.6.4
Precipitate, 4.2.17

INDEX

Precipitation, 4.2.18
    from homogeneous solution, 4.2.19
Precipitation indicator, 8.17.12
Precipitation titration, 8.27.12
Precision, 1.3.2, 2.3.11, 17.2.6, 18.4.3.2
    of a balance, 2.2.13, 2.3.12
    of indication, 2.2.03, 2.3.13
    of a weighing, 2.2.15
Precision balances, 2
Precision null-point potentiometry, 19.4.2
Precursor (or progenitor) ion, 7.46
Premix burner (in flame spectroscopy), 18.3.1.1.3
Presentation of results, 1
Pressure-gradient correction-factor, 13.6.05, 14.9.11
Primary (ion-selective) electrode, 21.2.1
Primary combustion (in flame spectroscopy), 18.3.1.1.3
Primary standard solution, 8.23.01
Primary substance, 8.24.01
Program (v.), 5.2.10
Programme (v.), 5.2.10
Programmed-current chronopotentiometry, 19.5.3
Prolate trochoidal mass spectrometer, 7.21
Protective agents, 18.4.4.3
Protonation constant, 11.3.4.2
Protonation equilibria, 11.3.4
Pulse polarography, 19.5.22
    derivative, 19.5.23
    differential, 19.6.3
Pulsed discharge lamp, 18.3.1.3.3.3

Quantity δ, 10.4.5
Quantity of a substance, 16.2.2, 17.2.1, 18.4.2
Quantum efficiency of fluorescence, 18.5.1.3, T. 18.5
Quasi-monochromatic photometry, 16.8.4.2
Quenching (in flames), 18.5.1.3

R_B value, 14.9.31
R_f value, 14.9.30
Radial electrostatic field analyser, 7.09
Radiance, spectral, 16.4.2, 16.6.3, T.18.5
Radiant energy, 16.4, 16.6.3, T.18.5
Radiant energy density, T.16.4, T.18.5
Radiant emissivity, T.16.4
Radiant exposure, T.16.4, 16.6.3, 16.8.2.2
Radiant flux, 16.4.5 (b), T.18.5
Radiant intensity, T.16.4, 16.6.3, T.18.5
Radiant quantity, 16.4.5
Radiative equilibrium, 18.6.1
Radioactive indicator, 8.19

INDEX

Radiofrequency (rf) polarography, 19.6.11

Radiometric end-point, 8.12.10

Range, 1.4.03

Range of applicability of a balance, 2.2.14, 2.3.14

Rate of liquid aspiration (in flame spectroscopy), 18.3.1.2

Rate of liquid consumption (in flame spectroscopy), 18.3.1.2

Rate of nucleation, 4.2.16

Ratio, volume swelling, 15.2.27

Readability, 2.2.11, 2.3.15

Reading, (in atomic spectroscopy), 18.3.5
    mg equivalent of, 2.2.12, 2.3.09

Real absorption, 18.5.1.2

Rearranged molecular ion, 7.44

Rearrangement, McLafferty, 7.66

Rearrangement ion, 7.49

Reciprocal linear dispersion, 16.5.2.4

Recovery factor, 12.07

Recovery, percentage, 1.4.11

Recovery test, 18.4.3.2

Redox indicator, 8.17.11

Redox ion-exchangers, 15.2.39

Redox polymers, 15.2.38

Redox titrations, 8.27.10

Reference beam, 18.3.3

Reference electrode, 19.1.8, 19.1.9, 21.1.09
    internal, 21.1.10

Reference element, 16.6.2.2, 16.8.4.4, 18.4.4.3

Reference intensity, 16.8.4.4

Reference solution, 18.4.1
    for pH*values in amphiprotic solutions, 10.7, 10.8

Reference-element technique, 18.4.4.3

Reflection factor, 16.5.3

Reflux nebulizer, 18.3.1.1.2

Refractive index, 16.5.2

Relative retention time, 14.9.28, 13.6.01, 15.2.40

Relative standard deviation, 1.4.06, 1.6.05, 17.2.4, 18.4.3.2

Releasers, 18.4.4.3

Reliability of results, 1.3

Reprecipitation, 4.2.36

Resin matrix, 15.2.09

Resolution, peak, 13.8.02, 14.9.37, 15.2.47
    practical, 16.5.2.2
    10 per cent valley definition, 7.23

Resolving power, 16.5.2.2
    in mass spectrometry, 7.24

Resonance fluorescence, 18.5.1.3

Resonance line, 18.5.1.1

Resonance spectrometer, 18.3.3

Response time, T.18.3.2, 18.3.5
Responsivity, 18.3.4
Rest point, 2.2.08, 2.3.16
Result, 1.2.2
    presentation of, 1
Retention index, 14.9.42
Retention parameters, 13.6, 13.7
Retention, relative, 13.6.08, 14.9.28, 15.2.40
Retention temperature, 14.9.29
Retention time, 13.6.01
Retention volume, 13.6.01, 13.6.05, 14.9.22
    adjusted, 13.6.03, 14.9.25, 15.2.41
    corrected, 13.6.04
    net, 13.6.06
    specific, 13.6.07, 14.9.27
    theoretical, 13.6.05, 13.9.04
    total, 14.9.23
Reversal dip, 18.5.1.2
Reversed direct-injection burner, 18.3.1.1.3
Reversed-phase chromatography, 14.6.8
Rigid matrix (ion-selective) electrode, 21.2.1.3
Ripening, Ostwald, 4.2.38
Rise velocity (of flame gases), T.18.3.1
Rotating-disc-electrode chronopotentiometry, 19.1.2
Rulings, total number of, 16.5.1.9

Salt form, of ion-exchanger, 15.2.14
Salting out, 12.17
Salting-out chromatography, 14.6.3
Sample, 3.1, 16.6.1, 18.4.1
Sample beam, 18.3.3
Sample injector, 13.3.01, 14.8.14
Sample size, 3.1
Sample weight classification, 3.2
Sampling boat, 18.3.1.3.1
Sampling cup, 18.3.1.3.1
Sampling loop, 18.3.1.3.1
Sand equation, 20
Saturated solution, 4.2.06
Saturation, 4.2.07
    chamber, 14.8.16
Saturation plateau, 18.4.4.3
Saturator, 18.4.4.3
Scale expansion, 18.3.5
Scale, zero point of, 2.2.09, 2.3.24
Scales of working in microchemical analysis, 3
Scatter, measurement of, 18.4.3.1
Scattering, of radiation, 18.5.1.2
Scavenger (collector), 4.2.22

INDEX

Scope of compendium, 0.4
Screened indicator, 8.17.13
Scrubbing, 12.15
Scrubbing solution, 12.16
Sealed lamp, 18.3.2
Second-derivative potentiometric titration, 19.4.7
Secondary combustion zone, 18.3.1.1.3
Secondary electron multiplier, 7.36
Secondary standard solution, 8.23.02
Secondary standard substance, 8.24.02
Secondary standards, for measurement of pH, 9.4
Selective elution, 14.6.4
Selectivity coefficient, 15.2.28
    corrected, 15.2.29
    potentiometric, 21.1.15
Self-absorption, 16.7.6.2, 18.5.1.2
Self-reversal, 16.7.6.3, 18.5.1.2
Semi-integral polarography, 19.1.8
Semimicro, 3.2
Semimicro sample, 3.2
Sensitivity, direct probe, 7.62.03
    for a stated load, 2.2.10, 2.3.17
    inlet system, 7.62.02
    ion-source, 7.62.01
Sensitivity, spectrochemical, 17.2.2, 18.4.2
Separate solution, method of, 21.3.4.2
Separated flame, 18.3.1.1.3
Separation factor, 4.2.33, 14.9.36, 15.2.31
Separation temperature, 14.8.24
Series, 1.2.4, 1.5.02
    sum of, 1.5.04
Shielded flame, 18.3.1.1.3
Sign convention, anodic currents 20
    cathodic currents, 20
    electrochemical data, 20
Simulation technique, 18.4.4.3
Simultaneous techniques (in thermoanalysis), 6.1.05, 6.2.17
Single-beam system (in flame spectroscopy), 18.3.3
Single-electrode plasma, 18.3.1.3.2
Single-element lamp, 18.3.2
Single-focusing mass spectrometer, 7.05
Single-sweep oscillographic polarography, 19.5.16
Single-sweep polarography, 19.5.16
Slit, spectral, 16.5.1.2
Slit-height, spectral, 16.5.1.2
Slit-width, spectral, 16.5.1.2
Slot burner, 18.3.1.1.3
Solid, active, 13.4.04, 14.8.09

INDEX

Solid angle (of radiation), T.18.3.2
Solid solution (mixed crystal), 4.2.32
Solid support, 13.4.03, 14.8.11
Solid volume, 13.3.02, 14.8.11.1
Solubility, 4.2.09
Solubility equilibria, 11.4
Solubility product, 4.2.08
Solute, 4.2.04
Solute volatilization interference (in flame spectroscopy), 18.4.4.2.2
Solution, 4.2.02
    comparison, 8.07
    method of separate, 21.3.4.2
Solution equilibria, symbols for, 11
Solution, saturated, 4.2.06
    supersaturated, 4.2.10
Solvent, 4.2.03
    amphiprotic, 10.1
Solvent blank, 18.4.1
Solvent extraction, 12.01
Solvent front, 14.8.22
Solvent migration-distance, 14.8.23
Sorption, 15.2.33
Sorption isotherm, 15.2.34
Spark line, 16.7.6.1
Spark source ionization, 7.32
Spatial distribution interference, 18.4.4.2.2
Special and *ad hoc* symbols in solution equilibria, 11.2.3
Specific interference, 18.4.4.2.2
Specific retention volume, 13.6.07, 14.9.27
Spectral band, 18.5.1.1
Spectral band width, 16.5.2.2, T.18.3.2
Spectral continuum, 18.5.1.1
Spectral frequency, 16.4.3
Spectral interference, 18.4.4.2.1
Spectral line, 16.7.6.1, 18.5.1.1
    intensity, T.18.5
    source, 18.3.2
Spectral period, 16.4.3
Spectral radiance, 16.4.2, 16.6.3, T.18.5
Spectral radiant energy, T.18.5
Spectral response curve, 18.3.4
Spectrochemical analysis, 16, 17, 18
Spectrochemical buffers, 16.7.7.2, 18.4.4.3
Spectrochemical carrier, 16.7.7.5
Spectrograph, 16.5.1.1, 18.3.3
    mass, 7.01
    resonance type, 18.3.3
Spectrometry, T.18.1
    mass, 7.04

INDEX

Spectroscope, mass, 7.03

Spectroscopic terms, 16.7.3

Spectroscopy, atomic emission, 16

Spectrum, mass, 7.39

Spontaneous electrogravimetry, 19.4.24

Spontaneous emission coefficient (Einstein), T.18.5

Spot, 14.7.08

Spray chamber, 18.3.1.1.2

Sprayer, 18.3.1.1.1

Square-wave polarography, 19.6.5

Staircase polarography, 19.6.2

Standard (known) addition method, 21.1.16

Standard deviation, 1.4.04, 1.5.08, 1.6.03, 17.2.3, 18.4.3.1

Standard, internal, 14.7.17, 16.8.4.4

Standard observer, 16.4.5 (b)

Standard solution, 8.23
    primary, 8.23.01
    secondary, 8.23.02

Standard substance, primary, 8.24.01

Standard substance, secondary, 8.24.02

Standard (known) subtraction method, 21.1.17

Standardization, 8.22

Standardization of pH and related terminology, 9, 10

Standards for measurement of pH, 9.4, 10.8
    secondary, 9.4

Static field mass spectrometer, 7.19

Starting point (or line), 14.7.09

Stationary phase, 13.4.05, 14.8.08

Stationary-phase fraction, 14.9.06

Stationary-phase volume, 14.9.05

Statistical weight, of particles in excitation states, T.18.6

Step (on an integral chromatogram), 13.5.03, 14.7.15

Step height (on an integral chromatogram), 13.5.03, 14.7.16

Step sector, 16.8.4.3

Stepwise elution, 14.6.5

Stepwise formation constant, 11.3.2

Stepwise line fluorescence, 18.5.1.3

Stirred-mercury-pool amperometry, 19.4.15

Stoicheiometric point, 8.13

Stokes fluorescence, 18.5.1.3

Stray light, 16.6.3.5

Stripping, 12.13
    anodic, 19.1.4
    cathodic, 19.1.4

Stripping analysis, 19.1.4

Stripping solution, 12.14

Submicro analysis, 3.3

Sub-trace analysis, 3.3.02

INDEX

Suction nebulizer, 18.3.1.1.2

Sum, of series, 1.5.04

Supersaturated solution, 4.2.10

Supersaturation, 4.2.11

Support plate, 14.8.12

Suppressor, 18.4.4.3

Suprathermal chemiluminescence, 18.5.1.1

Surface, 4.2.01

Swelling in solvent, weight, 15.2.26

Swelling ratio, volume, 15.2.27

Symbols, used in chromatography, 14.10
    used in mass spectrometry, 7.67
    used in solution equilibria, 11

T-tubes, 18.3.1.3.1

Tailing, 14.7.13

Tast polarography, 19.5.15

Temperature, absolute, 18.6.1

Temperature effects (gc retention data), 13.7.01

Temperature, injection, 14.8.25
    retention, 14.9.29
    separation, 14.8.24

Temperature-programmed chromatography, 14.6.1

Temperatures, initial and final, 14.8.26

Tensammetry, 19.3.2

Terms, spectroscopic, 16.7.4

Theoretical end-point, 8.13

Theoretical plate, height equivalent to a, 14.9.40
    height equivalent to an effective, 14.9.41

Theoretical plate number, 13.8.01, 14.9.38
    effective, 14.9.39

Theoretical retention volume, 13.6.05, 13.9.04

Theoretical specific capacity, 15.2.21

Thermal analysis, 6.1.01, 6.2.01, 6.2.06, 6.3.01
    derivative differential, 6.2.10, 6.3.03.5
    differential, 6.2.09, 6.3.03.4
    multiple techniques in, 6.1.05, 6.2.17, 6.3.06
    nomenclature of, 6, 6.1.01
    terms in, 6.2

Thermal decomposition, 6.1.06

Thermal equilibrium, 18.6.1

Thermal ionization, 7.29

Thermal radiation, 18.5.1.1

Thermoanalysis, 6.2.01

Thermoanalytical (adj.), 6.1.01

Thermodynamic definitions, 11.2.1

Thermodynamic equilibrium, 18.6.1

Thermogram, 6.1.04

Thermography, 6.1.01, 6.2.01

Thermogravimetric analysis, 6.2.04
    isothermal, 6.2.03
Thermogravimetric curve, 6.1.04, 6.3.02.3
Thermogravimetric thermogram, 6.1.04
Thermogravimetry, 6.2.04, 6.3.02.3
    derivative, 6.2.05, 6.3.02.4
Thermometric end-point, 8.12.11
Thermovaporimetric analysis, 6.2.13
Thin-layer chromatography, 14.5.4
Three-slot burner, 18.3.1.1.3
Tilt (of photographic plate), 16.5.1.6
Time-constant, 18.3.5
Time-of-flight mass spectrometer, 7.16
Time, retention, 13.6.01
Time resolved spectroscopy, 16.7.6.4
Titrant, 8.25
Titration, 8.26
   acid-base, 8.27.01
   acidimetric, 8.27.02
   alkalimetric, 8.27.03
   amperometric, 19.4.18
      with two indicator electrodes, 19.4.19
   back- , 8.04
   biamperometric, 19.4.19
   bipotentiometric, 19.4.11
   blank, 8.05
   chelatometric, 8.27.04
   compleximetric, 8.27.05
   conductometric, 19.2.2
      high-frequency, 19.2.4
   control, 8.09
   controlled potential coulometric, 19.4.27
   coulometric, 8.27.06, 19.4.14
      controlled potential, 19.4.27
   derivative potentiometric, 19.4.5
   dielcometric, 19.2.5
   dielectrometric, 19.2.6
   differential potentiometric, 19.4.4
   high-frequency conductometric, 19.2.4
   indirect, 8.27.07
   inverse derivative potentiometric, 19.4.6
   iodimetric, 8.27.08
   level of, 8.20
   non-aqueous, 8.27.09
   null-current potentiometric, 19.4.3
   oxidation-reduction, 8.27.10
   polarographic, 19.4.18
   polarometric, 19.4.18

INDEX

Titration (continued)
- potentiometric, 19.4.3
  - controlled current, 19.4.10
  - derivative, 19.4.5
  - differential, 19.4.4
  - inverse derivative, 19.4.6
  - null-current, 19.4.3
  - second-derivative, 19.4.7
  - with two indicator electrodes, 19.4.11
  - zero-current, 19.4.3
- second-derivative potentiometric, 19.4.7
- precipitation, 8.27.12
- weight, 8.27.13
- zero-current potentiometric, 19.4.3

Titration error, 8.28
Titrations, types of, 8.27
Titre, 8.29
Titrimetric analysis, 8, 8.30
Titrimetric conversion factor, 8.31
Total density (in atomic spectroscopy), T.18.6
Total ion current, after mass analysis, 7.56.01
  before mass analysis, 7.56.02
Total quantum efficiency of fluorescence in spectroscopy, T.18.5
Total retention volume, 14.9.23
Trace, 3.3
Trace constituent, 3.3
Transformation, 16.8.3
Transformation constant, 16.8.3.4
Transition interval, 8.32
Transmission factor, 16.5.3, T.18.5
Transmittance, photographic, 16.8.2.3
Triangular-wave polarography, 19.5.18
  cyclic, 19.5.20
Transit time, T.18.3.1
Transition probability (spectroscopic), T.18.5
Transport interference, 18.4.4.2.2
Travel time, T.18.3.1
Triangular-wave voltammetry, 19.5.19
  cyclic, 19.5.21
Tungsten-filament lamp, 18.3.2
Turbidimetric end-point, 8.12.12
Turbulent flame, 18.3.1.1.3
Twin nebulizer, 18.3.1.1.2
Two-colour indicator, 8.17.02
Two-dimensional chromatography, 14.6.7

Ulbricht's sphere, 16.A.5.5
Ultramicro analysis, 3.3
Ultrasonic nebulizer, 18.3.1.1.2

INDEX

Ultra/trace analysis, 3.3.01
Untergrund, 16.6.3.3

Vacuum phototube, 18.3.4
Value of a division (in precision balances), 2.2.04, 2.3.18
Vapour phase interference, 18.4.4.2.2
Variance, 1.4.05, 1.5.07, 1.6.04, 17.2.5, 18.4.3.1
Variate, 1.2.3, 1.5.01
Variates, number of, 1.4.07
Velocity, interstitial, 14.9.20
Vibrating-reed electrometer, 7.37
Visual end-point, 8.12.13
Visual indicator, 8.16
Volatilization (in spectroscopy), 18.3.1.1.1
Volatilized fraction (local), 18.3.1.2
Volatilizer, 16.7.7.4, 18.4.4.3
Voltage, ionizing, 7.26
Voltammetry, anodic stripping, 19.1.4
    cyclic triangular-wave, 19.5.21
    derivative, 19.5.10
    differential, 19.5.11
    hydrodynamic, 19.5.9
    linear-sweep, 19.5.8
    stationary-electrode, 19.5.8
    triangular-wave, 19.5.19
        cyclic, 19.5.21
Volume, adjusted retention, 13.6.03, 14.9.25, 15.2.41
    bed, 14.9.02
Volume capacity, 15.2.22
Volume, column, 14.9.01, 15.2.19
    corrected retention, 13.6.04
    dead, 14.9.10
    designated, 8.10
Volume distribution coefficient, 15.2.30.3
Volume, gas hold-up, 13.6.02, 14.9.09
    hold-up, 14.9.08
    interstitial, 13.3.04, 14.9.03
    liquid, 13.3.03
    net retention, 13.6.06, 14.9.26
Volume of the stationary phase, 14.9.05
Volume, peak elution, 14.9.24
    retention, 13.6.01, 13.6.05, 14.9.22
    solid, 13.3.02, 14.8.11.1
    specific retention, 13.6.07, 14.9.27
    theoretical retention, 13.6.05, 13.9.04
    total retention, 14.9.23
Volume swelling ratio, 15.2.27
Volumetric flow rate, 14.9.18

INDEX

Wavelength, 16.4.2, T.18.5
    of maximum transmission, T.18.3.2
Wavenumber, 16.4.2
Weighting procedures, 2.3.19
Weight-change determination,
    isobaric, 6.2.02, 6.3.02.1
    isothermal, 6.2.03, 6.3.02.2
Weight swelling in solvent, 15.2.26
Weight titration, 8.27.13
Width, of light beam, 16.5.1.8
Width, of peak, 13.5.02, 14.9.16, 15.2.45

Xenon lamp, high-pressure, 18.3.2

Zero-current potentiometric titration, 19.4.3
Zero-current potentiometry, 19.4.1
Zero-point of scale, 2.2.09, 2.3.24
Zone (in chromatography), 14.7.07